Robert G. Seippel
Instructor, Electronics
College of the Canyons,
Valencia, California

Roger Lincoln Nelson
Development Engineer
Honeywell Corporation
Minneapolis, Minnesota

American Technical Society
Chicago 60637

Library of Congress Catalog Number: 74-24890
ISBN 0-8269-1589-2

123456789 75 987654321

Printed in the United States of America

Preface

The analysis of alternating current (ac) electrical circuits is almost an ancient art. Nearly everyone from Benjamin Franklin to Penelope Priscilla Pendergast has written a dissertation on the subject. In most cases these texts are outstanding, with circuit coverage complete but not always brief and simple. In other cases the texts are reflections of the writers' backgrounds, with compilations of a wealth of technical knowledge.

Great detail is understandably necessary in particular circumstances. There is a definite need and place for detail and unavoidable complexity, especially in research and science, and most engineering. No writer would attempt to compile any technical presentation without valid reference material supplied by the tireless effort of reliable and highly qualified specialists. But despite the thousands of papers written on the subject, instructors still find themselves searching for a simpler method for students to effectively analyze ac circuits.

The objective of this book is to provide this simpler method. Procedures outlined in the text have been constructed in cookbook style so they may be closely followed by the beginner and also provide a quick reference to the journeyman.

Art work is drawn in stages, to establish continuity. Some electrical basics are also covered but details are kept to "nutshell" descriptions.

Methods for using trigonometry tables or the slide rule for solving electrical problems are described in detail in the appendix. Operations of mathematics using powers of ten are also explained. The appendix also contains extensive tables of ac formulas for quick reference.

The writer acknowledges all others who have attempted an easier way and recognizes that this book could not possibly fill all the needs of everyone concerned with the details of ac circuit analysis. Hopefully, the simplified approach offered here can aid a better understanding of the most used laws and theories on the subject without requiring the reader to wade through swamps of complexity.

Robert G. Seippel

THANKS TO:

Hazel, my wife—Typist
Dal Fitts—Technical Editor

TABLE OF CONTENTS

Section *Page*

LIST OF ILLUSTRATIONS

Fig. No. *Page No.*

LIST OF TABLES

Section 1

Basic Facts About Alternating Current

Alternating current is much more versatile than direct current. The basic fact that it alternates makes it easily transferrable without much power loss, whereas direct current loses much power in transfer.

Alternating current is able to radiate at different time intervals (frequencies). This makes it useful in most electronic fields such as radio, radar, television, etc. About the only advantage that direct current has over alternating current is that direct current can be stored.

WHAT IS ALTERNATING CURRENT? (SEE FIG. 1)

Alternating current is electrical current that varies in magnitude and reverses in polarity. A pure sine wave of alternating current continuously varies in magnitude and periodically reverses direction. The electric charge (electrons) moves back and forth along the

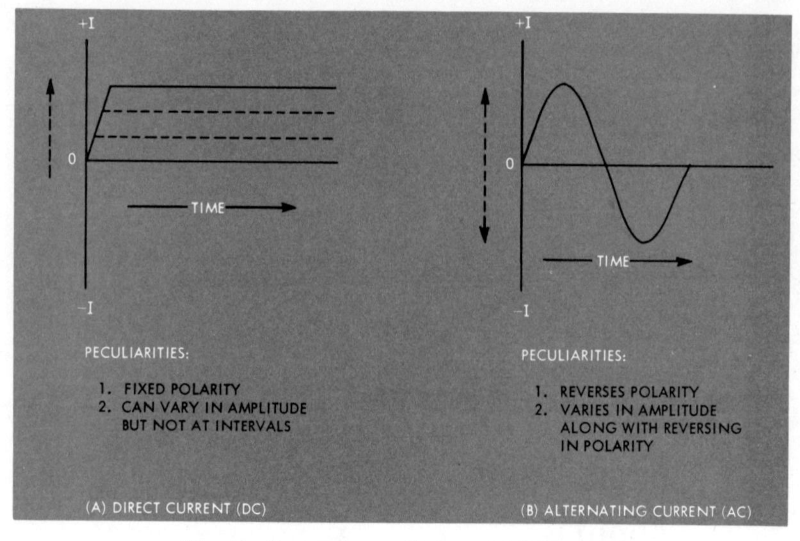

Fig. 1. Comparison of DC and AC Current

circuit like a weight on a string (pendulum) swings. Such motion is called "sinusoidal" or "simple harmonic."

Alternating currents are directly associated with alternating voltages, and thus the notation *ac* is commonly used to refer to both alternating current and alternating voltage. It is assumed at this time that the reader has knowledge of direct current (dc) circuit fundamentals. This assumption is necessary to preclude the problem of reviewing electron theory and dc electrical fundamentals. A comparison of direct current (dc) and alternating current (ac) is made in Fig. 1.

DEVELOPMENT OF AN AC SINE WAVE (SEE FIG. 2)

In Fig. 2, a loop of wire is rotated counterclockwise through a magnetic field. As the loop passes N (north), induced current is maximum because the loop is cutting the flux lines at the maximum rate. As the loop passes 180 degrees, induced current is minimum (zero) because the loop is not cutting the flux lines. Maximum induced current occurs again at 270 degrees but the polarity is changed to negative. It can easily be seen that an alternating current varies in magnitude and periodically changes direction of flow.

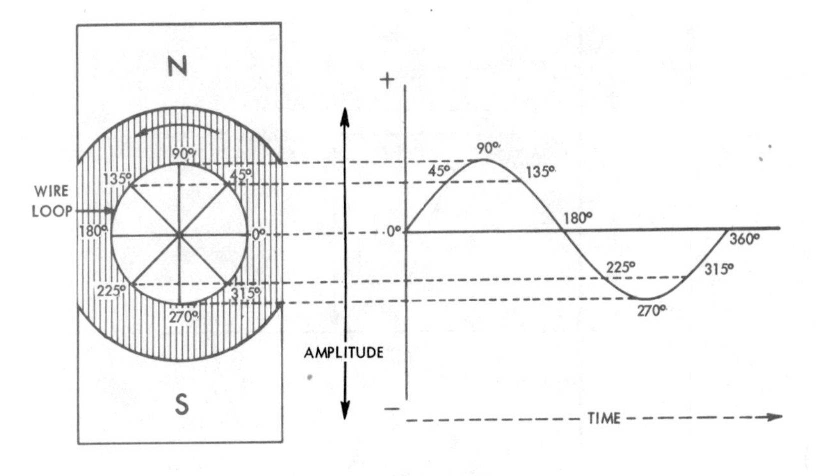

(A) SINE WAVE DEVELOPMENT

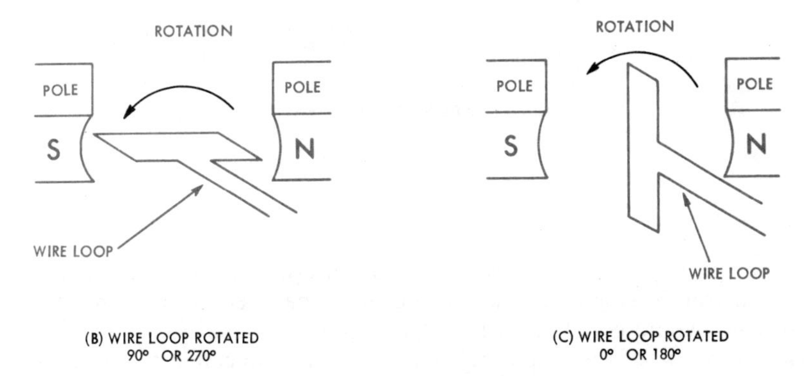

(B) WIRE LOOP ROTATED 90° OR 270°

(C) WIRE LOOP ROTATED 0° OR 180°

Fig. 2. Development of an AC Sine Wave

PARTS OF A SINE WAVE (SEE FIG. 3)

The parts of a sine wave are easily shown with the diagram in Fig. 3. One complete cycle is achieved in 360 degrees of rotation as shown in the development of a sine wave. A single alternation (half of a cycle) is completed at 180 degrees. Maximum positive amplitude to maximum negative amplitude is peak to peak (voltage or current). Maximum amplitude to minimum amplitude (zero) is peak (voltage or current). Average current or voltage is 0.636 × peak and effective (rms) is 0.707 × peak.

An instantaneous voltage is at any point on the sine waves and is the voltage existing at any one instant of time. Average voltage

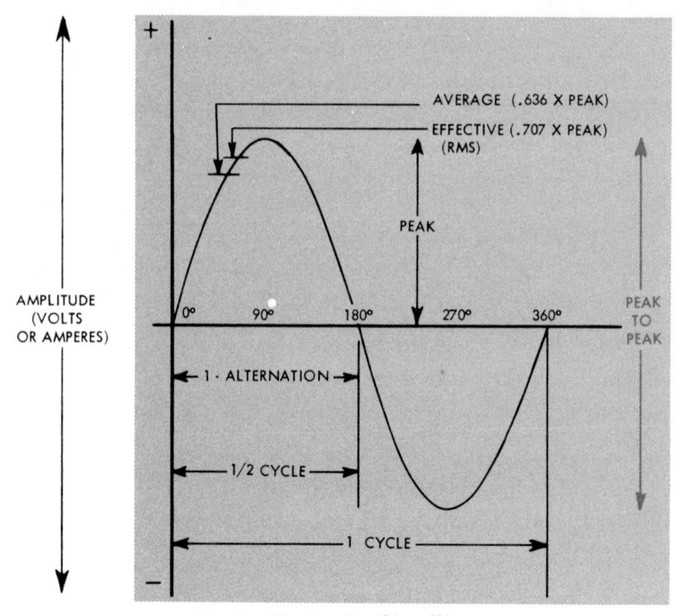

Fig. 3. Parts of a Sine Wave

is the average of all the points of an alternation. This average is taken from a single alternation, because the average of both alternations would equal zero. Effective voltage is equal to the trigonometric value of the sine of 45 degrees. It is also called rms or root-mean-square value (used in commercial power line calculations).

Conversion from one value of ac measurement to another is a simple matter of multiplication. Conversion is often necessary in maintenance and trouble-shooting processes. Table 1 illustrates the measurement and how to convert it to some other ac measurement.

TABLE 1. AC MEASUREMENT CONVERSION

CONVERSION FROM	MULTIPLICATION FACTOR			
	RMS	AVERAGE	PEAK	PEAK TO PEAK
RMS (EFFECTIVE)	1.00	0.900	1.414	2.828
AVERAGE	1.110	1.000	1.570	3.141
PEAK	0.707	0.636	1.000	2.000
PEAK-TO-PEAK	0.354	0.318	0.500	1.000

FREQUENCY (SEE FIG. 4)

The number of cycles generated in a second is called the frequency. One cycle per second is shown in Fig. 4(A). Note that the entire cycle is within the 1 second time interval. If the frequency were 60 cycles per second (hertz), there would be 60 complete cycles within the 1 second time interval. In Fig. 4(B), two waves are compared. Note that wave *A* is 2 cycles within the one second time interval and wave *B* is 4 cycles within the 1 second time interval. Since wave *B* has more cycles per second than wave *A*, its time interval per cycle is smaller.

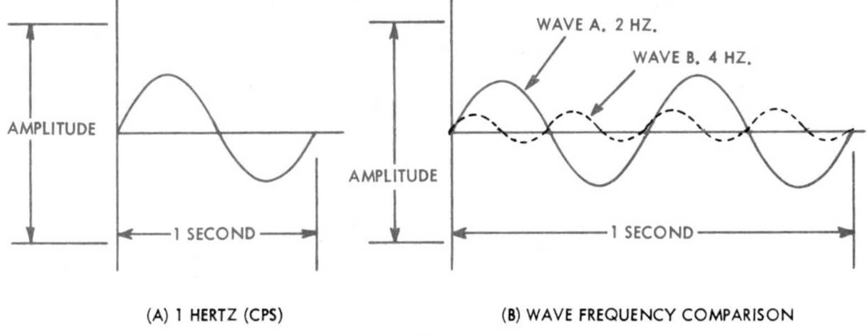

Fig. 4. Frequency

FREQUENCY SPECTRUM (SEE TABLE 2)

AC frequencies range from much less than 1 cycle per second (hertz) to millions of cycles per second (megahertz) and further to billions of cycles per second (gigahertz). The spectrum of frequency has been provided with short names to be used for different jobs. These ranges are segregated into chart form called a *frequency spectrum*. In Table 2 a typical spectrum is provided with typical applications.

WAVELENGTH

Wavelength is defined as the length along a line of wave propagation that is occupied by one cycle. The formula for wavelength is

$$\lambda(\text{lambda}) = \frac{\text{V(velocity of propagation)}}{\text{f(frequency of wave)}} = \text{T(period)} = \frac{1}{\text{f(frequency)}}$$

Three forms of this formula are used in the same manner as Ohm's law.

TABLE 2 FREQUENCY SPECTRUM

RANGE	USE AND APPLICATION
O HERTZ	Direct current (DC) power application.
16 TO 16K HERTZ	Audio (A-F) frequencies. Sounds that can be heard by man.
10K HERTZ TO 30M HERTZ	Radio (VLF, LF, MF, HF) frequencies. Above man's hearing level. Man can hear up to 17K hertz. Used for radio communications, navigation, international short wave, etc.
30M HERTZ TO 300M HERTZ	Very high radio frequency (VHF). Used for radio, navigation, FM broadcast band, television channels 2 through 13 etc.
300M HERTZ TO 3000M HERTZ	Ultra high radio frequency (UHF). Used for radio navigation, radar, television channels 14 to 83, etc.
3,000M HERTZ TO 30,000M HERTZ	Super high radio frequencies (SHF). Used in microwaves wavelength 1-10 cm.
30G TO 300G HERTZ	Extra high radio frequencies (EHF). Used in microwave wavelengths 0.1-1cm.
K = 1000 (KILO) M = 1,000,000 (MEGA) G = 1,000,000,000 (GIGA)	

WHERE:

(1) $\lambda = \frac{V}{f}$ — λ = Wavelength, usually in meters

(2) $f = \frac{V}{\lambda}$ — V = Velocity of propagation

(3) $V = f\lambda = \frac{\lambda}{T}$ — T = Period of 1 cycle

f = Frequency of wave in cycles per second (hertz)

Velocity of sound waves in air is 1,130 feet per second and should be used in the formula for V when calculating sound waves.

PERIOD

The time for completion of 1 cycle is called the period. Period and frequency are the inverse, or reciprocal, of each other. That is

60 hertz and 1/60 of a second; also, 1000 hertz and 0.001 second, or 1 millisecond.

PROPAGATION (SEE FIG. 5)

When voltage is generated at one point (actually between two points) in a circuit, it does not appear instantly throughout the circuit but most propagate thru the circuit. The rate at which it travels depends on the characteristics of the circuit.

In Fig. 5, we see an ac generator connected to a pair of wires (say, TV lead-in wire); at the other end we see a resistor. Due to the propagation time the voltage sine wave across the resistor is somewhat behind the sine wave at the generator. Let's assume the frequency is high enough so that the propagation time is 1/4 of a period. We would find then that the voltage sine wave across the resistor would be 90° behind the generator.

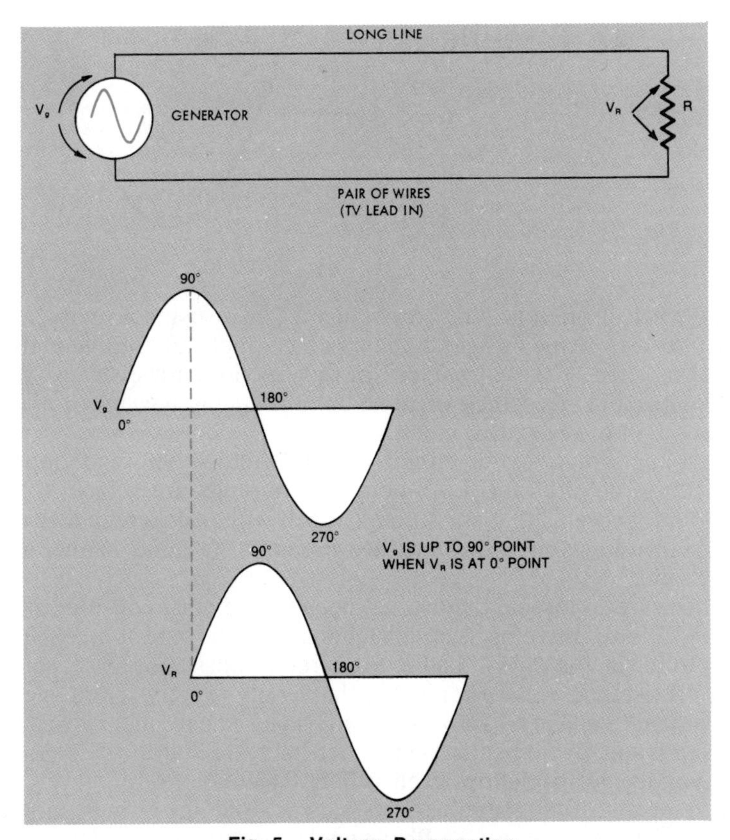

Fig. 5. Voltage Propagation

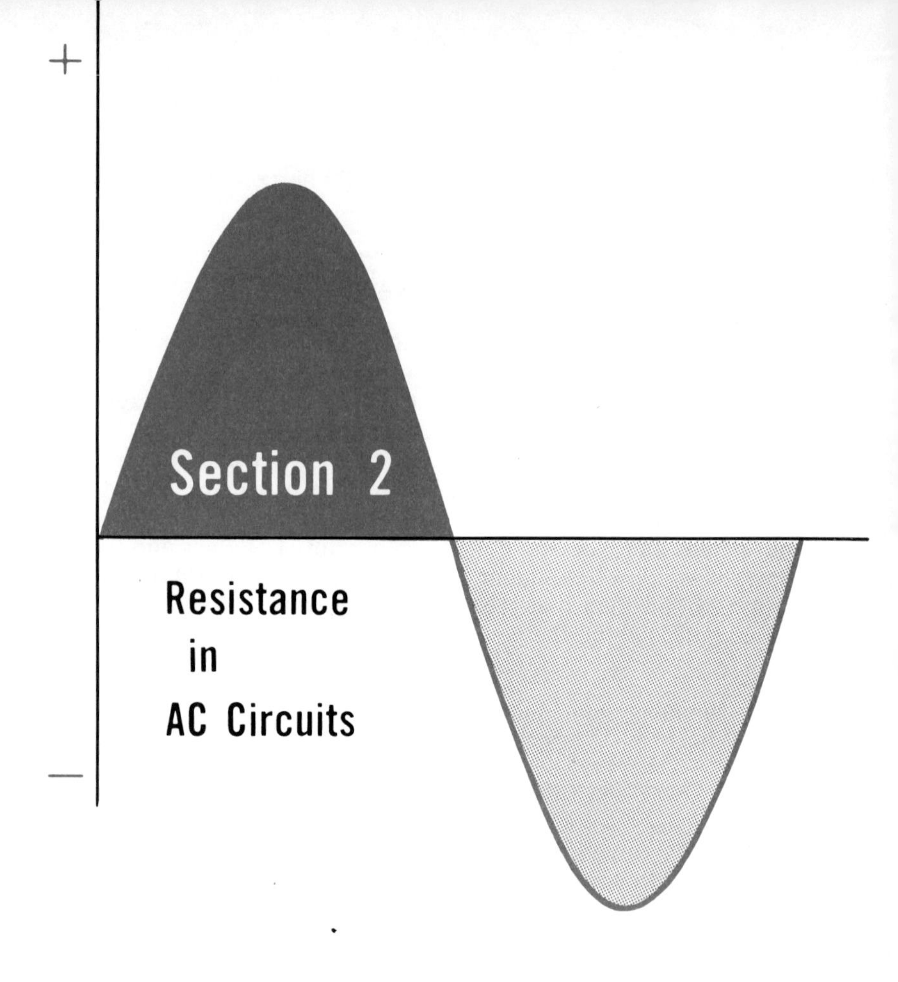

Section 2

Resistance in AC Circuits

All electrical circuits have resistance. The resistance may be in many forms, such as a light bulb or a power tool and, for that matter, the resistance of the wire itself. In this Section, however, we shall deal with pure resistance in an ac circuit, simply because it makes explanation of resistance much simpler.

It is imperative that dc circuitry be understood before attempting to analyze ac circuitry. Both ac and dc circuits are subject to the laws and theorems that are normally dealt with in dc circuit analysis. Resistance in ac circuits is handled in much the same manner as in dc circuits.

There are exceptions, however, because we must consider things such as phase, waveforms and frequency. Let us take as an example the circuit in Fig. 6(A). The entire circuit consists of an ac source and a resistance. In ac circuits that have only resistive components, current and voltage waveforms are in phase. The frequency we have chosen is one cycle per second (1 hertz). Note that the current and the voltage forms follow each other. Calculations for purely resistive ac circuit are simple. Ohm's law is used just as in dc circuits with resistance (Fig. 6(C).) Phase angles are not a consideration in

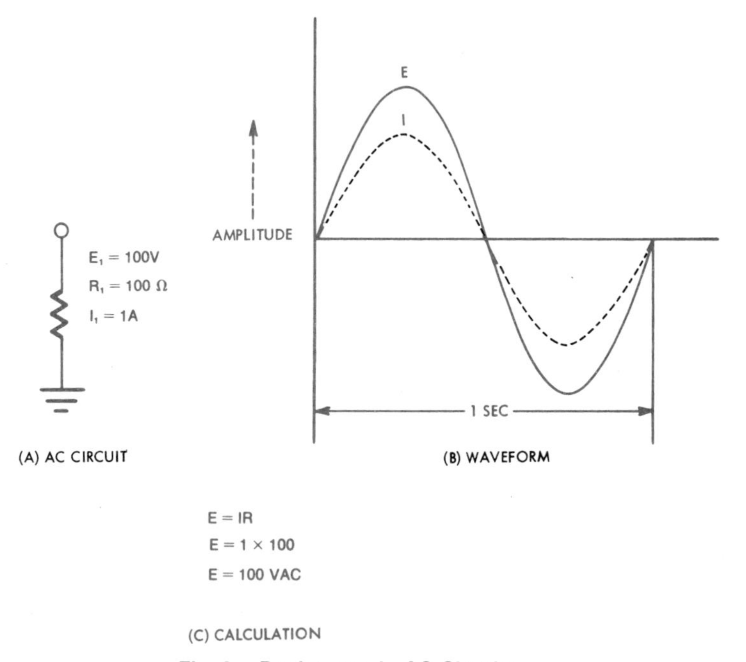

Fig. 6. Resistance in AC Circuits

that there is no difference in phase (0 degrees) between current and voltage.

Later, when considering ac circuits that have inductance and capacitance, the phase angle must be considered. Measurement for ac circuits with resistance is in terms of either peak, effective, or average voltages. If you measure voltage in peak values then current also should be handled in peak values. Effective and average values should be handled with the same general thinking.

RESISTORS IN SERIES AC CIRCUITS (SEE FIG. 7.)

Pure resistive ac series circuits are analyzed in the same manner as their dc series circuit counterparts. The three rules for analyzing dc series circuits apply: (a) Total voltage is the sum of the voltage drops ($E_T = E_1 + E_2 +$ etc); (b) Total resistance is the sum of the resistances ($R_T = R_1 + R_2 +$ etc); and (c) Current is the same throughout the circuit ($I_T = I_1 = I_2 =$ etc).

All the current and voltage phase relationships are in phase as shown in Fig. 7(D). This statement is true for all values of voltages

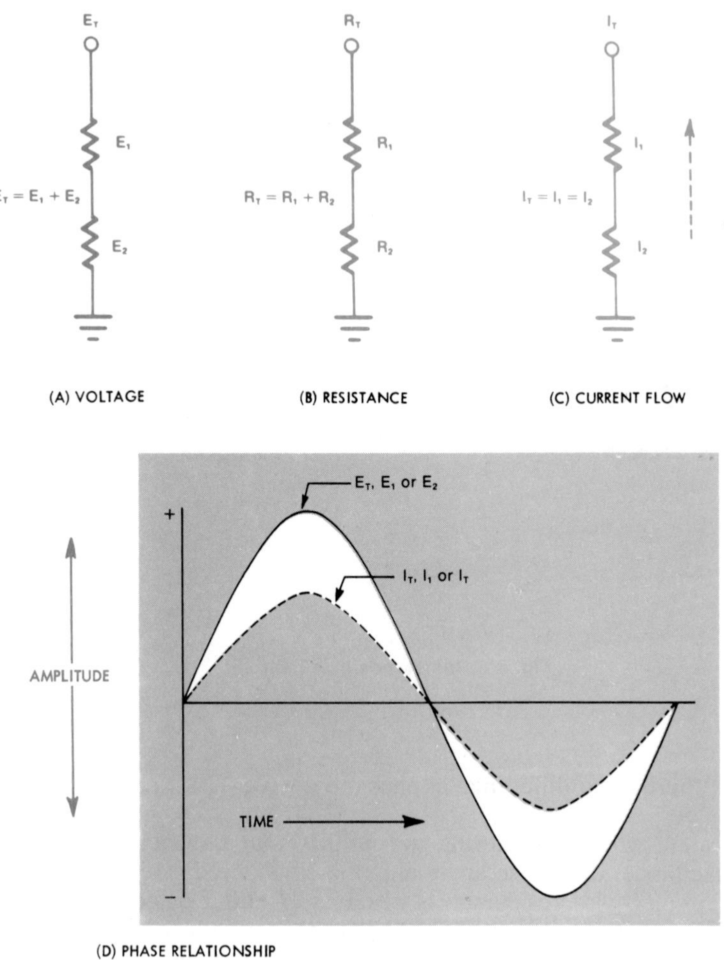

Fig. 7. Resistors in Series AC Circuits

or currents in the circuit. Circuits that have only resistance are named *pure resistive*, and have a zero phase relation between voltage and current.

Calculations for total voltage are as follows:

GIVEN:

$E_1 = 10$ volts

$E_2 = 5$ volts

FIND: E_T

$E_T = E_1 + E_2$

$E_T = 10 + 5$

$E_T = 15$ volts

Calculations for total resistance are as follows:

GIVEN:

$R_1 = 10$ ohms

$R_2 = 15$ ohms

FIND: R_T

$R_T = R_1 + R_2$

$R_T = 10 + 15$

$R_T = 25$ ohms

Calculations for total current are as follows:

GIVEN:

$I_1 = 10$ amps

$I_2 = 10$ amps

FIND: I_T

$I_T = I_1 = I_2$

$I_T = 10$ amps

RESISTORS IN PARALLEL AC CIRCUITS (SEE FIG. 8)

Pure resistive ac parallel circuits are analyzed in the same manner as their dc parallel circuit counterparts. The three rules for analyzing dc parallel circuits apply: (a) The same voltage is applied to all parallel legs ($E_T = E_1 = E_2 =$ etc); (b) Resistance total may be calculated by the reciprocal formula, $\left(\frac{1}{R_T} = \frac{1}{R_1} + \frac{1}{R_T} + \text{etc}\right)$; and (c) Total current is the sum of the currents in all branches ($I_T = I_1 + I_2 +$ etc).

In pure resistive ac circuits, all currents and voltages are in phase as shown in Fig. 8(D). This statement is true for all values of voltages or currents in the circuit. Circuits that have only resistance are

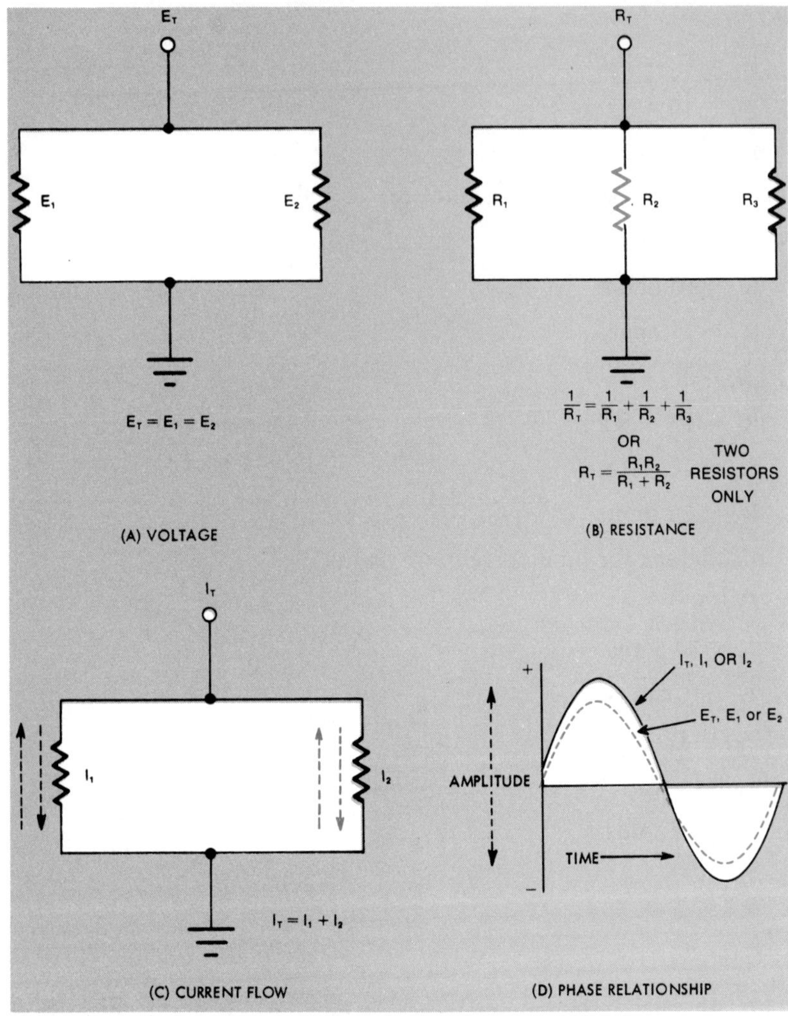

Fig. 8. Resistors in Parallel AC Circuits

named *pure resistive* and have a zero phase relation between voltage and current.

Calculations for total voltage are as follows:

GIVEN:

$E_1 = 10$ volts

$E_2 = 10$ volts

FIND: E_T

$E_T = E_1 = E_2$

$E_T = 10$ volts

Calculations for total current are as follows:

GIVEN:

$I_1 = 10$ amps

$I_2 = 5$ amps

FIND: I_T

$I_T = I_1 + I_2$

$I_T = 10 + 5$

$I_T = 15$ amps

Calculations for total resistance are as follows:

GIVEN:

$R_1 = 10$ ohms

$R_2 = 20$ ohms

$R_3 = 20$ ohms

$$\frac{1}{R_T} = \frac{1}{R_1} + \frac{1}{R_2} + \frac{1}{R_3}$$

$$\frac{1}{R_T} = \frac{1}{10} + \frac{1}{20} + \frac{1}{20}$$

$$\frac{1}{R_T} = \frac{2}{20} + \frac{1}{20} + \frac{1}{20}$$

$$R_T = \frac{20}{4} = 5 \text{ ohms}$$

Calculations for total resistance in circuits with only two resistances are as follows:

GIVEN:

$R_1 = 60$ ohms

$R_2 = 20$ ohms

FIND: R_T

$$R_T = \frac{R_1 R_2}{R_1 + R_2}$$

$$R_T = \frac{60 \times 20}{60 + 20}$$

$$R_T = \frac{1200}{80}$$

$$R_T = 15 \text{ ohms}$$

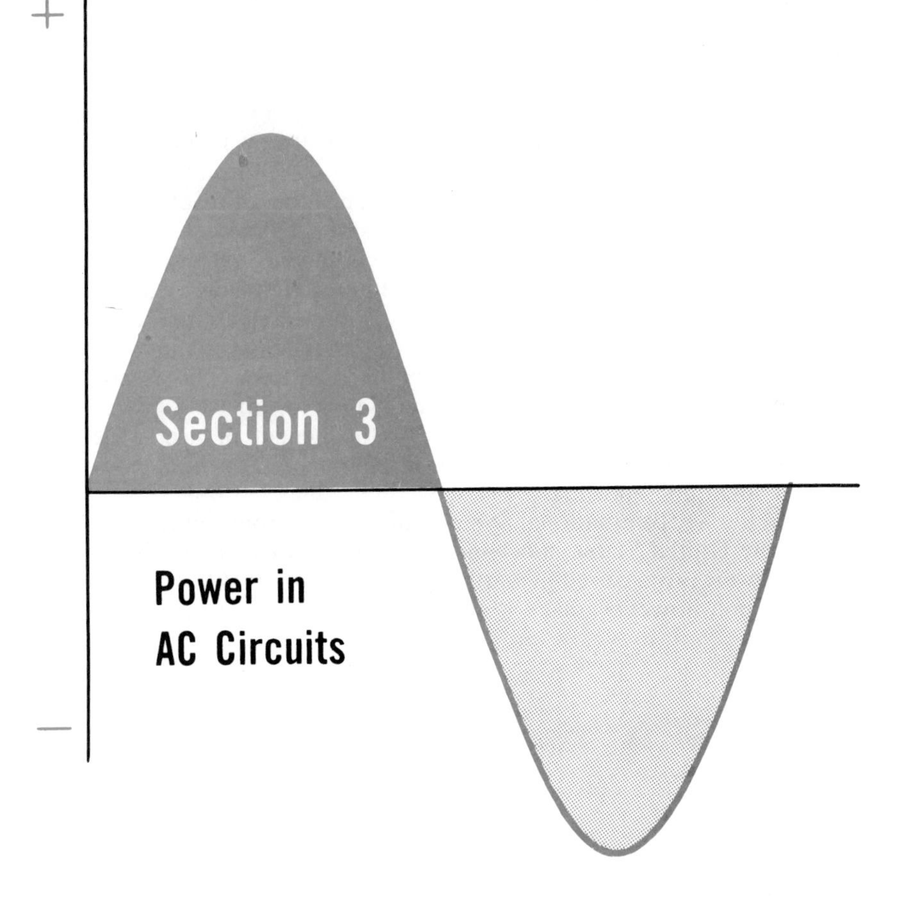

Section 3

Power in AC Circuits

There are many power formulas for ac circuits. These formulas are listed in Appendix *C* in the back of this book. There are some points concerning power that should be cleared up before we concern ourselves with formulas, however.

Real power can be calculated using the total resistance in the circuit. Real power is measured in *watts*. The formula $P = I^2R$ is used. In the event voltage and current are used in a circuit with reactance and resistance, real power is calculated using the formula $P = EI \cos \angle\theta$. This is required because of the phase angle between current and voltage. Either formula can be used depending on which is more suitable for the situation.

The power factor in a series ac circuit is the ratio of resistance to impedance or $PF = \frac{R}{Z}$. In a parallel circuit the power factor is the ratio of resistive current to total current or $PF = \frac{I_R}{I_T}$.

Apparent power is that power in a circuit which has no resistance. It is calculated by multiplying voltage times current. Since current and voltage are 90 degrees out of phase in pure reactance, power is only apparent. *Real power* is measured in watts (W). Since no power is actually dissipated in pure reactance, the formula P = EI is measured in *volt-amperes* (VA) when speaking of reactive circuits.

Currently (1975) the United States still uses the customary U.S. system of measurements for most technologies except electricity, where the units are metric, such as amperes, volts, ohms, and watts. But in some applications, particularly in power electricity, electrical units have to be translated or converted to mechanical units such as *horsepower* (HP) in the customary U.S. System. This is possible due to the fact that 746 watts or 0.746 kilowatt (real power) are equivalent to one horsepower. For the benefit of users, electric motors in the United States are therefore rated in horsepower or fractional horsepower.

In the SI (International System) of metrics proposed for adoption by the United States, mechanical power, as well as all other forms of power, is measured in watts (W) or kilowatts (kW). Knowing this does not, however, make it an accomplished fact at this time.

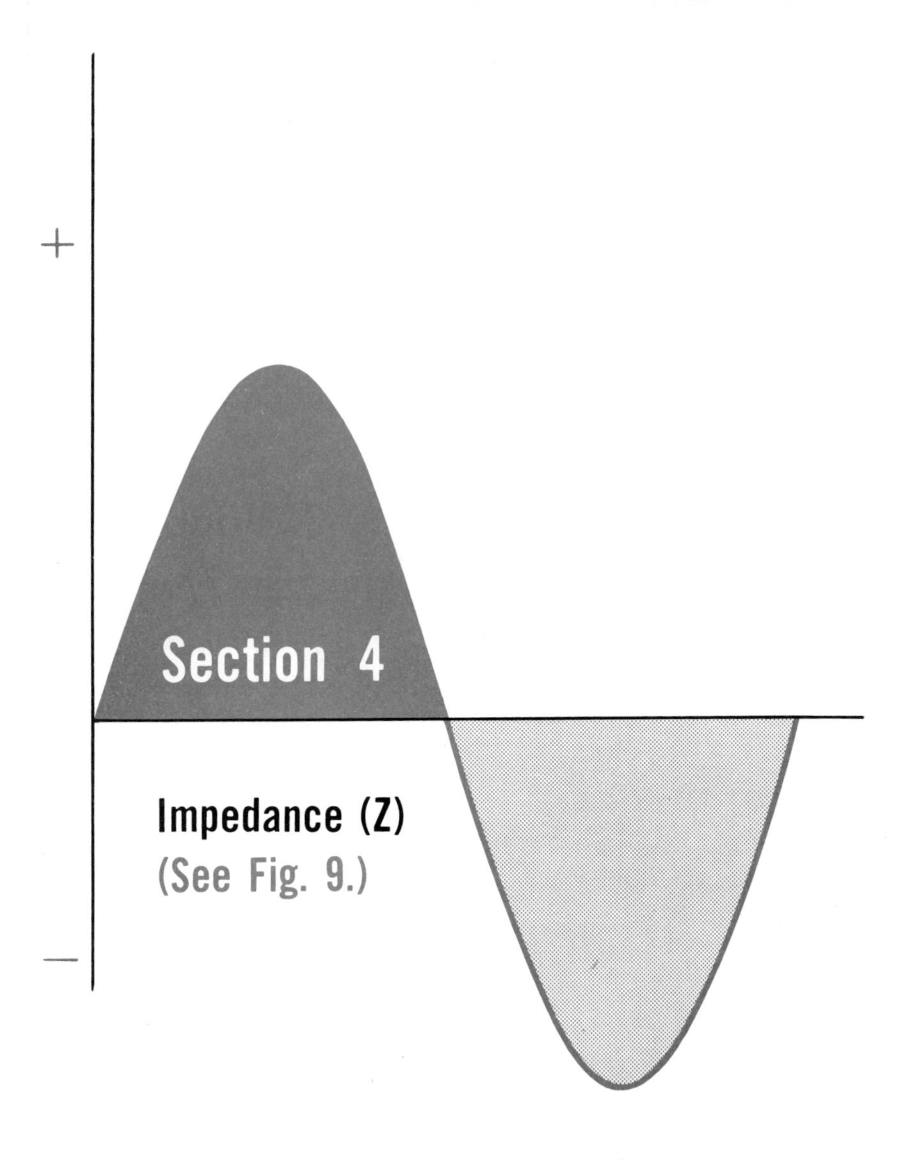

Section 4

Impedance (Z)

(See Fig. 9.)

Impedance is defined as total opposition to current flow in an ac circuit.

In a purely resistive circuit, the impedance is the same as the total resistance of the circuit. In a purely capacitive circuit the impedance is the same as the total capacitive reactance of the circuit. In a purely inductive circuit, the impedance is the same as the total inductive reactance of the circuit.

In mixed circuits, such as resistive-capacitive (RC), resistive-inductive (RL) and resistive-capacitive-inductive (RCL) circuits, total impedance (Z) is found with the use of the Pythagorean theorem and Ohm's law. Analysis of impedance in mixed circuits will be covered in later sections under RC, RL, and RCL circuit analysis.

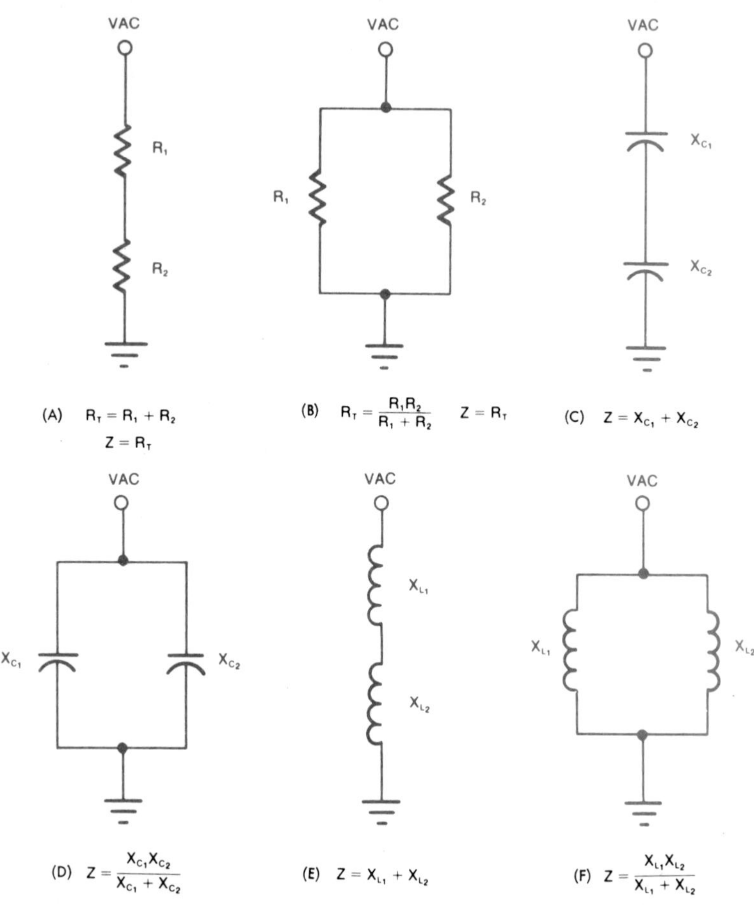

Fig. 9. Impedance in Pure Resistive, Capacitive, and Inductive Circuits

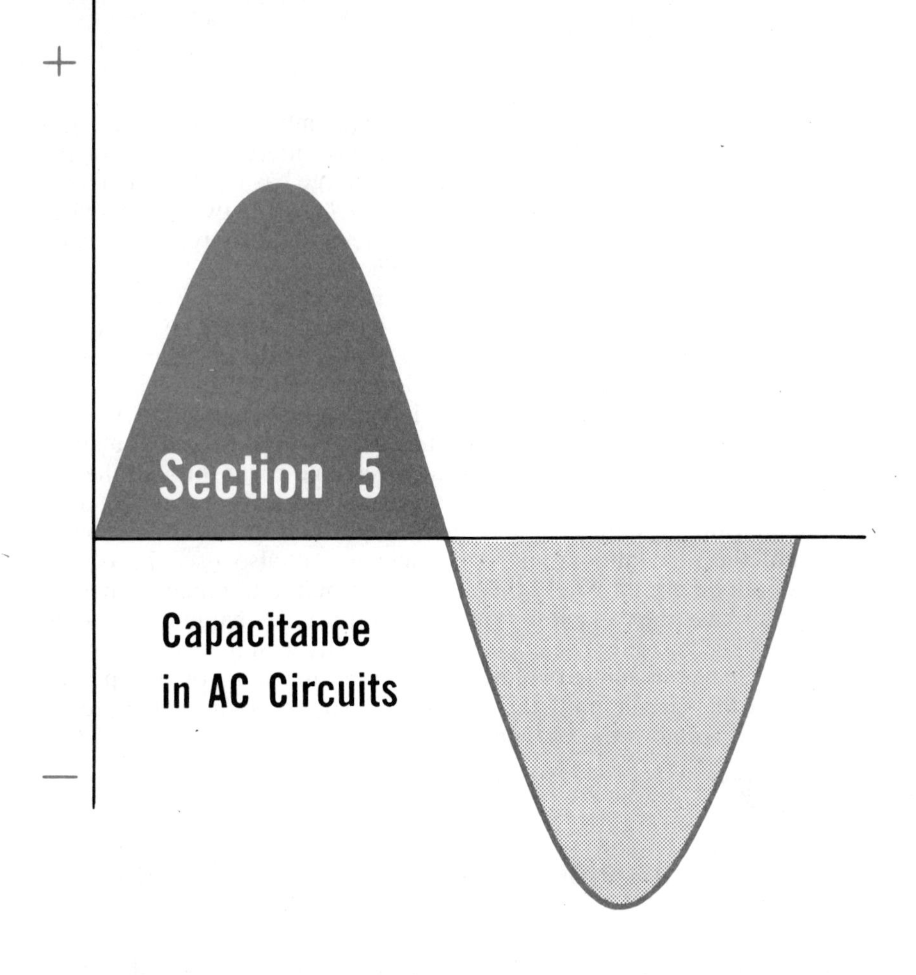

AC circuits use three basic components. These are resistors, capacitors, and inductors. The capacitor which we are dealing with in this Section is also called a *reactor*. Its basic purpose in ac circuits is to react to (oppose) a change in voltage. By its reaction, voltage can be used for electronic sine wave propagation and signal development.

THE CAPACITOR

The capacitor is a device for storing an electrical charge. It is made of two conductors (plates) that are separated by a *dielectric* insulator. When an electrical source is applied to the capacitor an excess of electrons produces a negative charge on one side of the capacitor. An equal positive charge on the other side of the capacitor denotes a lack of electrons. Note that the capacitor *as a whole* does not have an excess nor deficiency of electrons. Thus, when we say that a capacitor has a charge of 1 coulomb on it, our meaning is that one coulomb of negative charge has entered one side of the capacitor and 1 coulomb of electronic charge has left the other side.

The charge causes an *electrostatic field* (F) between the plates which is measured in volts per centimeter. The charge remains on the capacitor after the power source is disconnected.

To discharge the capacitor, a circuit is connected across the capacitor and the excess electrons on its negative side flow into and through the external circuit back to its positive side, where there is a lack of electrons, and the two sides are neutralized.

This action is tricky because, in reality, current is not flowing *through* the capacitor as it appears to. It is merely stored as a static charge on the capacitor plates, then released to go through the external circuit when the voltage drops toward zero on each alternation. Now the static charge is on the opposite plate, where it remains until the next alternation. This capacitor is also used in a pure, smooth dc circuit where the voltage is required to remain constant. In this case the capacitor would build up a static charge which could not be discharged. The effect would be equivalent to an open circuit. Any change in applied voltage across the capacitor would be passed on to the circuit until the capacitor charges or discharges to an amount equal to the change.

CAPACITOR DETAIL AND THE FARAD AS A UNIT OF MEASURE

Plate area, dielectric thickness, dielectric strength, and dielectric constant determine the capacitance of a capacitor (in farads).

$C = k \times \frac{A}{d} \times 22.4 \times 10^{-14}$ is the mathematical formula for capacitance (where C = capacitance in farads, k is the dielectric constant, A is the area of either plate in sq. inches, d is the distance between plates in inches, and 22.4×10^{-14} is the conversion to farads).

In actual use, a charge (Q) of one coulomb is equal to one volt potential (E) across a one-farad capacitor (C) $Q = CE$. Refer to charge factor (Q) under a separate heading. In practical use the capacitor is in sizes which range between microfarads (μF) and picofards (pF).

The word *microfarad* means one-millionth farad ($10^{-6}F$), and the word *picofarad* means one-trillionth farad ($10^{-12}F$). These very small sub-multiples of the farad are required for electronics, where the farad is much too large a unit for practical use. Use of these metric prefixes and powers of ten are fully explained in Appendix B: Calculating with Powers of Ten.

CAPACITOR SYMBOLS (SEE FIG. 10)

Capacitor symbols vary with manufacture. However, the standard capacitor symbols according to military specifications are shown in the Fig. 10. These are also recognized throughout industry.

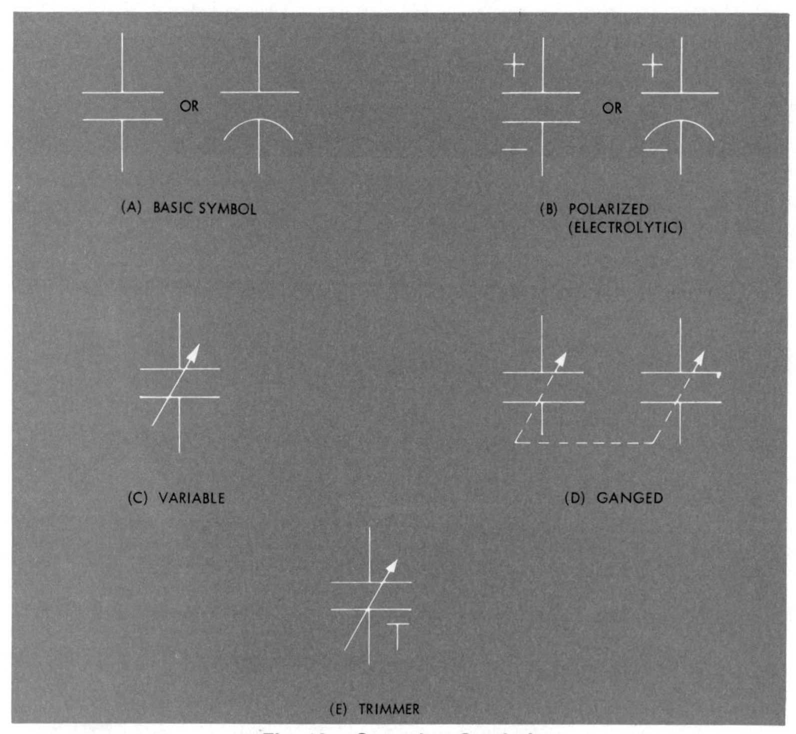

Fig. 10. Capacitor Symbols

CAPACITOR TYPES (SEE FIG. 11)

Capacitor dielectrics are made from many types of material. The most commonly used are air, ceramic, mica, paper and electrolytics. Different types are chosen because of their size, the job to be done, breakdown volts, ability to withstand heat or vibration, cost, and service life. Reference to vendor manuals will provide construction data and allow the reader to choose the right capacitor for the job.

The air capacitor is usually a variable capacitor. The most common is the tuning capacitor for radio receivers. A fixed set of plates and a movable set of plates are used to change the capacitance. Air is the dielectric. The plates must not touch one another.

Ceramic capacitors are made from baked earth. They are usually the disc or tubular type. Silver plates are fused on opposite sides of the ceramic dielectric to be used as the plates.

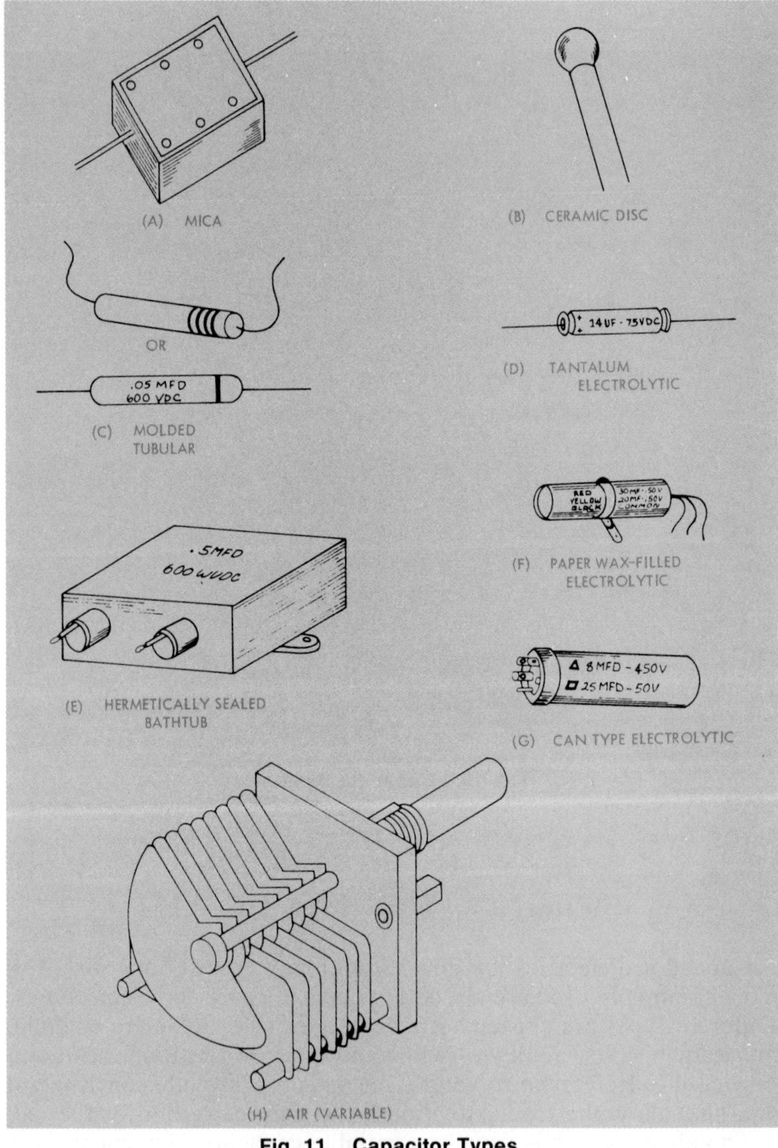

Fig. 11. Capacitor Types

Mica capacitors are made by placing mica sheets between tinfoil, then baking the sandwich in a phenolic case. The tinfoil provides the conductor plates and the mica the dielectric.

Paper capacitors are made by wrapping tinfoil and paper into a roll. The roll is then placed in a cardboard cylinder and sealed with wax. The tinfoil provides the conductor plates and the paper is the dielectric.

Tantalum capacitors are an electrolytic type used in low-voltage applications. Electrodes (plates) are separated by a gauze saturated with an electrolyte. An oxide film is formed on the plates and provides a layer between the positive plates and the gauze. The oxide is formed by applying direct current to the capacitor in the correct polarity, and this oxide film is an insulator. If current of enough magnitude is applied in the opposite direction, the oxide film will break down and the capacitor will be destroyed. The materials described are all placed in a compact cylinder. Aluminum electrolytic capacitors are made in the same manner as tantalum capacitors. In fact, aluminum was used before tantalum in the construction of electrolytic capacitors.

CAPACITOR COLOR CODING

Tubular and mica capacitors are sometimes color coded. Other capacitors have their farad values and power ratings written on their outer coverings.

Color coding for mica capacitors is slowly being replaced by printed values on capacitor covers. The reason for this is that there is confusion between the RMA, EIA, MIL, JAN, and AWS methods for color coding. Since confusion exists it may be difficult for the user to determine which color code to use. See Fig. 12 for color coding of mica capacitors. See Fig. 13 for color coding of tubular capacitors. See Fig. 14 for obsolete (but still in use) capacitor color coding.

Some critics might object to including an illustration of obsolete color coding here. This is a case where merely declaring that a design, product, or method is obsolete does not make it so. It still exists until all uses of it have ceased. All conditions that a student is likely to encounter are not ideal, and for that reason a really practical book should prepare the student to deal with present situations.

CHARGE FACTOR (Q) OF A CAPACITOR

The charge (Q) of a capacitor is the charge stored on the capaci-

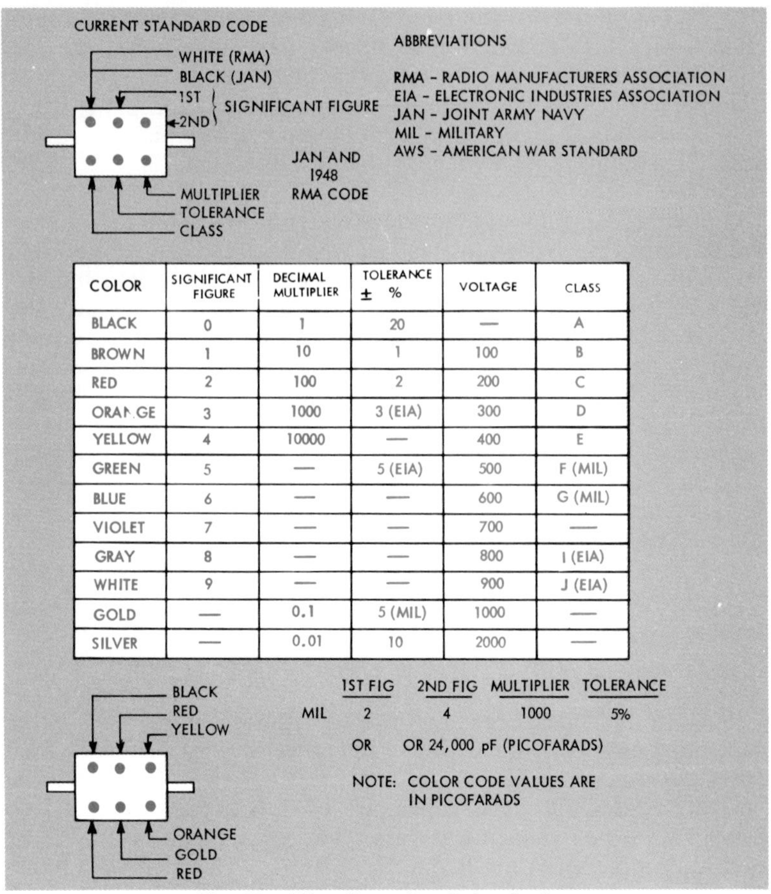

COLOR	SIGNIFICANT FIGURE	DECIMAL MULTIPLIER	TOLERANCE ± %	VOLTAGE	CLASS
BLACK	0	1	20	—	A
BROWN	1	10	1	100	B
RED	2	100	2	200	C
ORANGE	3	1000	3 (EIA)	300	D
YELLOW	4	10000	—	400	E
GREEN	5	—	5 (EIA)	500	F (MIL)
BLUE	6	—	—	600	G (MIL)
VIOLET	7	—	—	700	—
GRAY	8	—	—	800	I (EIA)
WHITE	9	—	—	900	J (EIA)
GOLD	—	0.1	5 (MIL)	1000	—
SILVER	—	0.01	10	2000	—

Fig. 12. Mica Capacitor Color Coding

tor's plates. The greater the charging voltage (E), the greater the charge. Also, the greater the capacitance (C) with a set voltage, the greater the charge amount.

These relationships of voltage and capacitance can be analyzed in the formula $Q = CE$, where Q is the charge in coulombs, C is the capacitance in farads, and E is the applied voltage in volts.

When 1 volt is applied to a capacitor with a value of 1 farad, the charge on the capacitor is 1 coulomb, ($Q = 1 \times 1$). If a larger capacitor, say of 5 farads, is substituted, the charge would be 5 coulombs ($Q = 5 \times 1$). Since the formula $Q = CE$ is a direct proportion, it is obvious that the size of the capacitor or applied voltage directly affects the charge Q.

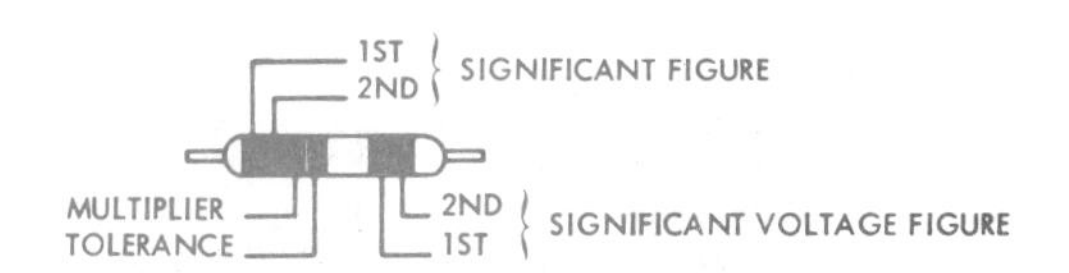

COLOR	SIGNIFICANT FIGURE	DECIMAL MULTIPLIER	TOLERANCE	VOLTAGE VOLTS
BLACK	0	1	20	—
BROWN	1	10	—	100
RED	2	100	—	200
ORANGE	3	1000	30	300
YELLOW	4	10000	40	400
GREEN	5	100,000	5	500
BLUE	6	1,000,000	—	600
VIOLET	7	—	—	700
GREY	8	—	—	800
WHITE	9	—	10	900
GOLD	—	0.1	—	—
SILVER	—	—	10	—
NO COLOR	—	—	—	—

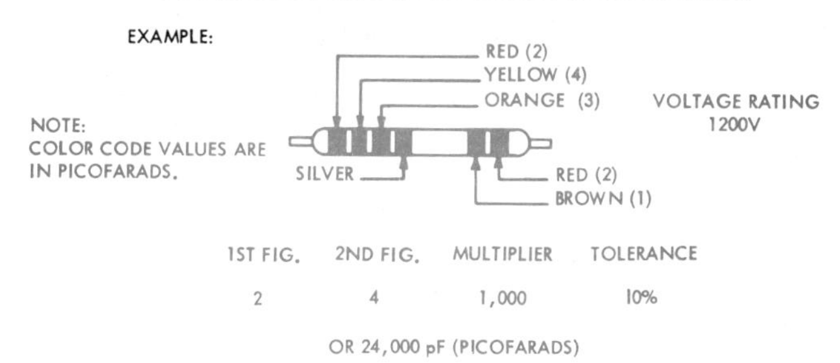

Fig. 13. Tubular Capacitor Color Coding

CAPACITANCE IN SERIES AC CIRCUITS (SEE FIG. 15)

Capacitance in series ac circuits are added using the reciprocal formula, in the same manner as resistors in parallel. In Fig. 15 note that there are two formulas. The first provides the reciprocal formula to be used for more than two capacitors. Solution is as follows:

GIVEN: *NOTE:* μF = microfarads

$C_1 = 10\ \mu F$

$C_2 = 20\ \mu F$

$C_3 = 20\ \mu F$

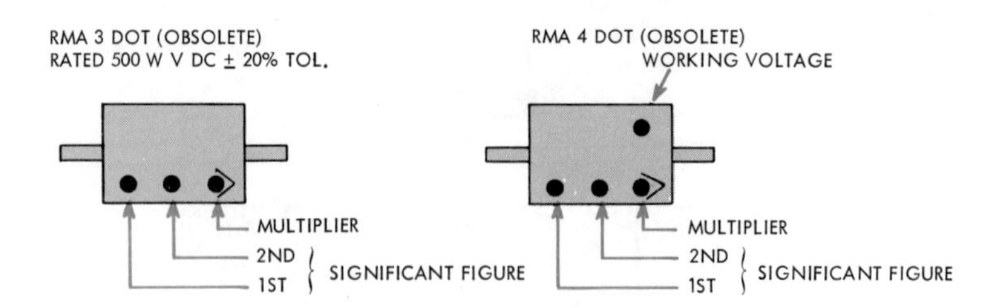

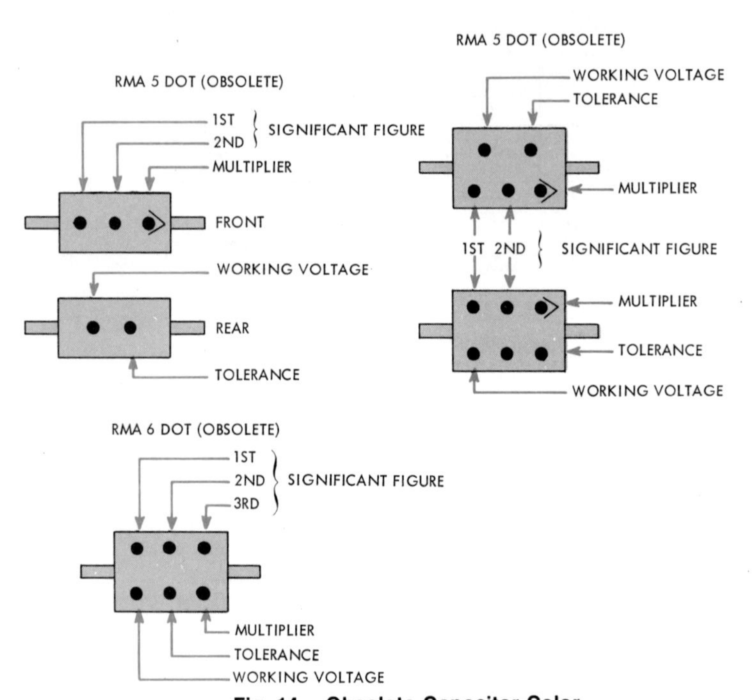

Fig. 14. Obsolete Capacitor Color

FIND: C_T

$$\frac{1}{C_T} = \frac{1}{C_1} + \frac{1}{C_2} + \frac{1}{C_3}$$

$$\frac{1}{C_T} = \frac{1}{10} + \frac{1}{20} + \frac{1}{20}$$

$$\frac{1}{C_T} = \frac{2}{20} + \frac{1}{20} + \frac{1}{20} = \frac{4}{20}$$

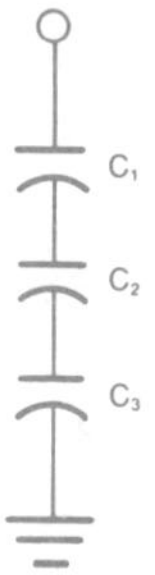

C_1

C_2

C_3

$$\frac{1}{C_T} = \frac{1}{C_1} + \frac{1}{C_2} + \frac{1}{C_3}$$

OR

$$C_T = \frac{C_1 C_2}{C_1 + C_2} \quad \left\{ \begin{array}{l} \text{FOR TWO} \\ \text{CAPACITORS} \end{array} \right\}$$

(A) CIRCUIT (B) FORMULA

Fig. 15. Capacitors in Series AC Circuits

$$C_T = \frac{20}{4} = 5\ \mu F$$

Calculations for total capacitance in series ac circuits with only two capacitors is as follows:

GIVEN: *NOTE:* μF = microfarads

$C_1 = 10\ \mu F$

$C_2 = 20\ \mu F$

FIND: C_T

$$C_T = \frac{C_1 C_2}{C_1 + C_2}$$

$$C_T = \frac{10 \times 20}{10 + 20}$$

$$C_T = \frac{200}{30}$$

$$C_T = 6.6\ \mu F$$

CAPACITOR VOLTAGE DIVIDERS (SEE FIG. 16)

Capacitors may be used as voltage dividers. The reactance of an ac capacitive voltage divider is inverse to that of a dc capacitor voltage divider.

Let's first take a look at the ac voltage divider in the Fig. 16(A). When the capacitances are converted to capacitive reactance the

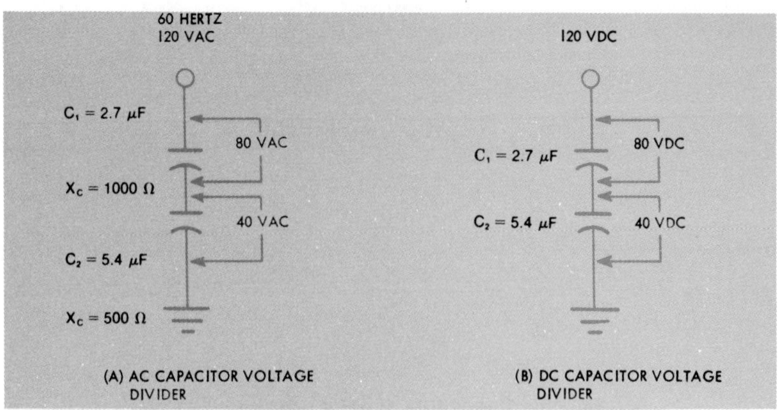

Fig. 16. AC Versus DC Capacitor Voltage Division

voltage drops are as in a series resistive circuit (additive). Therefore the ratio of the reactances is two to one and the voltage drop across capacitor C_1 is 80 vac and C_2 is 40 vac.

In the dc voltage divider in Fig. 16(B), the capacitors have the same charging current. The capacitors C_1 and C_2 therefore have the same charge. The voltage drop across each is inversely proportional to the capacitance. Since this is the case $\left(E = \frac{Q}{C}\right)$, then the capacitor C_2 being twice the size of capacitor C_1 would have one half the voltage drop.

CAPACITORS IN PARALLEL AC CIRCUITS (SEE FIG. 17)

Capacitors in parallel ac circuits are added in the same manner as resistors in series. Solution is as follows:

GIVEN: *NOTE:* μF = microfarads

$C_1 = 10\ \mu F$

$C_2 = 20\ \mu F$

$C_3 = 20\ \mu F$

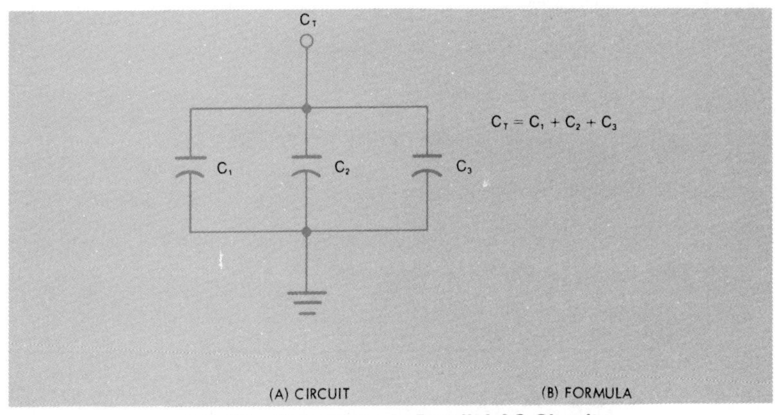

Fig. 17. Capacitors in Parallel AC Circuits

FIND: C_T

$$C_T = C_1 + C_2 + C_3$$

$$C_T = 10 + 20 + 20$$

$$C_T = 50\ \mu F$$

Section 6

Capacitive Reactance in AC Circuits

Capacitive reactance is opposition to current flow in a capacitive ac circuit. It is calculated in ohms the same as resistance; however, it cannot be measured with an ohmmeter (only calculated). The reason for this is that a capacitor responds to *changes* in current with time.

The capacitive opposition to current flow is a ratio of voltage to current in an ac circuit. Capacitive reactance is distinguished from pure resistance by the symbol X_C. Capacitive reactance is affected by two variables: capacitance of the capacitor (in farads), and the frequency of the ac current flowing in the circuit (in hertz).

For pure sinusoidal (sine wave) circuits, this opposition to current flow (capacitive reactance) may be calculated with the following formula:

$$X_C = \frac{1}{2\pi fC}$$

Where X_C = Capacitive reactance
π = 3.1416 (constant)
f = Frequency of the current in hertz
C = Capacitance of the capacitor in farads

It may be noted that an increase of frequency or capacitance will decrease capacitive reactance. Likewise a decrease in frequency or capacitance will increase capacitive reactance. This increase or decrease is non-linear.

Low-frequency ac currents applied to a capacitor will create high reactance (approaching infinity). High-frequency ac currents applied to a capacitor will create low reactance. The same is true for an increase or decrease in capacitance. Fig. 18 illustrates capacitive reactance in relation to a change in frequency or capacitance.

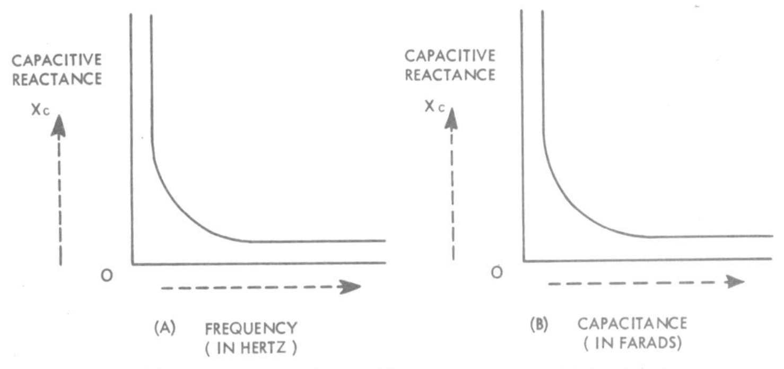

Fig. 18. Change in Capacitive Reactance As Frequency or Capacitance Vary

PURE CAPACITIVE PHASE ANGLES (SEE FIG. 19)

In a pure capacitive ac circuit, the current sine wave leads the voltage sine wave by 90 degrees at any instantaneous point in time. It could also be said that under these circumstances voltage lags current.

This is explained simply: When the current is at 90 degrees in time, voltage is at 0 degrees in time; when current is at 180 degrees in time, voltage is at 90 degrees in time, etc. At the 0 and 180 degree position in time, current changes direction. At the 90 and 270 degree position in time, voltage changes direction. However, the directions are opposite.

As the current increases in amplitude, voltage is decreasing in amplitude and vice versa. With pure capacitance, this is true. Inserting resistance in the circuit causes the phase relationship of voltage and current to vary. This change will be explained under RC series and parallel circuit analysis.

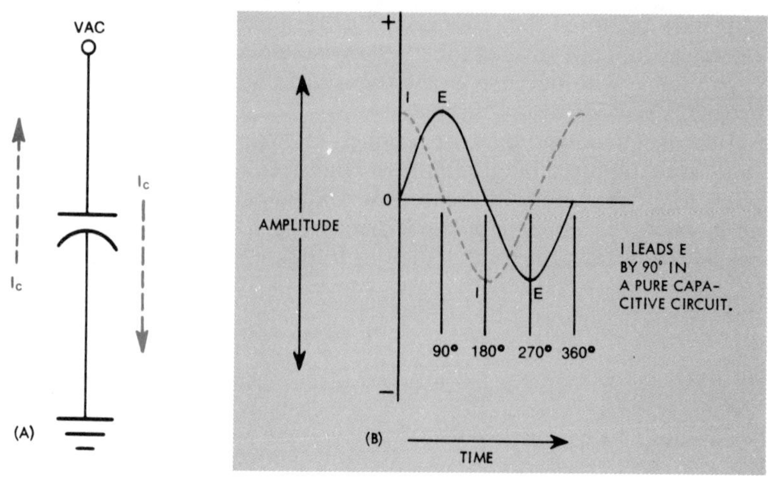

Fig. 19. Pure Capacitive Phase Angles

RC TIME CONSTANT (SEE FIG. 20)

The time constant in a resistive-capacitive circuit is actually the time in seconds that it takes for the capacitor in the circuit to charge to 63.2 percent of its final charge.

An RC circuit contains at least one capacitor and one resistor. Fig. 20 illustrates this circuit with a voltage input of 5 volts. When the switch is closed, current begins to charge the capacitor. The capacitor charges by 63.2 percent of applied voltage in the first time constant.

In the second time constant, the charge voltage across the capacitor increases by 63.2 percent of the remaining applied voltage, as follows:

APPLIED VOLTAGE	5.00 volts
1st TIME CONSTANT	− 3.16 volts
REMAINING VOLTAGE	1.84 volts

In the 2nd time constant voltage will increase 63.2 percent of 1.84 volts: $0.632 \times 1.84 = 1.16$ volts increase.

The charge voltage on the 3rd, 4th, and 5th time constants increase in the same manner. After the 5th time constant, for practical application, the charge has reached the full 5.0 volts applied.

While the charge is building up from 0 to 5 volts the capacitor is opposing a change in voltage. The reaction of the RC circuit during this time is called *transient response*. The time constant (TC) in seconds is a ratio of capacitance and resistance as shown below.

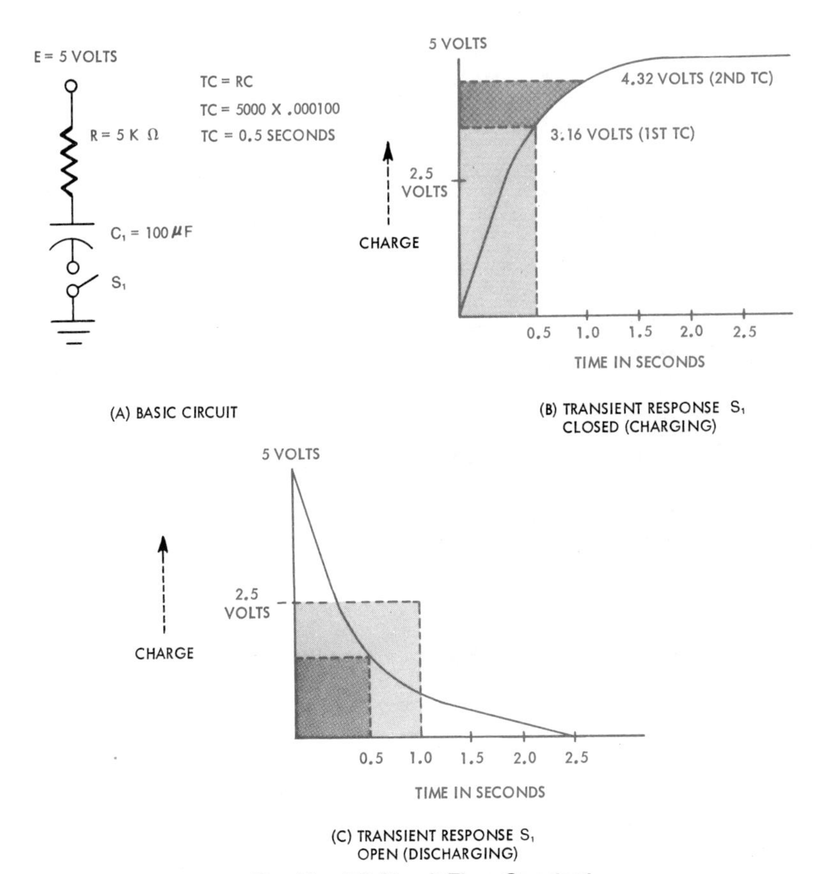

Fig. 20. RC Circuit Time Constant

$$TC = RC$$

$$TC = 5000 \times 0.0001 \qquad (5\ k\Omega \times 100\ \mu F)$$

$$TC = 0.5\ \text{seconds}$$

If the closed circuit shown in Fig. 20(A) is now opened, the reaction of the circuit will be that the capacitor will discharge through the resistor by 63.2 percent in the first time constant, then by 63.2 percent of the remaining charge in the second time constant. The capacitor will continue to discharge in the same manner, and after the fifth time constant will, practically speaking, have a zero charge.

The time constant grows in value if either the capacitance or the resistance is increased. Conversely, then, the time constant becomes less if either the capacitance or the resistance is decreased.

Section 7

Series Resistive-Capacitive (RC) Circuit Analysis

In a series RC circuit, resistance and capacitance are used together to develop the sine wave. The most significant factor in the ac series RC circuit is that the voltage lags the current. This is useful in control devices and other circuits where time delays are required.

TOTAL VOLTAGE IN A SERIES RESISTIVE-CAPACITIVE (RC) CIRCUIT (SEE FIG. 21)

In a series RC circuit, total voltage is calculated using the root-mean-square (rms) equation based on the Pythagorean theorem (vector sum). This theorem is stated electrically as $E_T{}^2 = E_{X_C}{}^2 + E_R{}^2$ or $E_T = \sqrt{E_{X_C}{}^2 + E_R{}^2}$.

In Fig. 21, the horizontal side of the triangle is plotted as the volt-

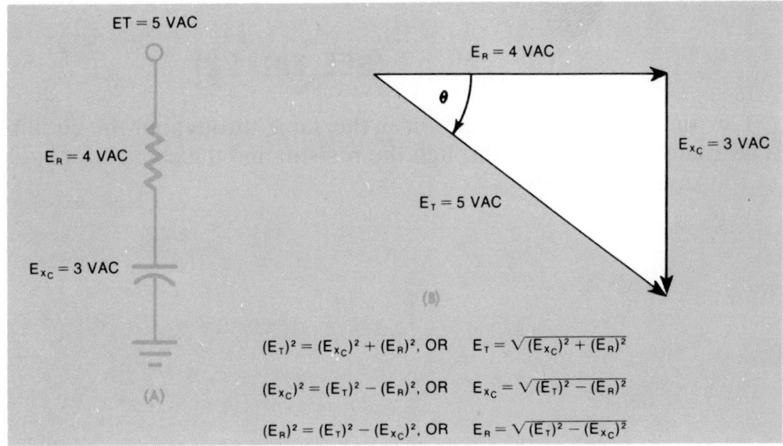

Fig. 21. Total Voltage in a Series RC Circuit

age drop across the resistance (E_R). The vertical side, which is negative in direction, is plotted as the voltage drop across the capacitor (E_{X_C}). The reason for this is that the X_C voltage lags current in a capacitive circuit. The angle, being negative, is drawn in a clockwise direction. The plotted angle shown in the figure is sometimes called a *phasor*. Calculation procedures are as follows:

① *IF E_T IS UNKNOWN* *THEN:* $E_T^2 = E_{X_C}^2 + E_R^2$

GIVEN:

$E_{X_C} = 3$ volts

$E_R = 4$ volts

$$E_T^2 = 3^2 + 4^2$$
$$E_T = \sqrt{9 + 16}$$
$$E_T = \sqrt{25}$$
$$E_T = 5 \text{ volts}$$

② *IF E_{X_C} IS UNKNOWN* *THEN:* $E_{X_C}^2 = E_T^2 - E_R^2$

GIVEN:

$E_T = 5$ volts

$E_R = 4$ volts

$$E_{X_C}^2 = 5^2 - 4^2$$
$$E_{X_C} = \sqrt{25 - 16}$$
$$E_{X_C} = \sqrt{9}$$
$$E_{X_C} = 3 \text{ volts}$$

③ *IF E_R IS UNKNOWN* *THEN:* $E_R^2 = E_T^2 - E_{X_C}^2$

GIVEN:

$E_T = 5$ volts

$E_{X_C} = 3$ volts

$$E_R^2 = 5^2 - 3^2$$
$$E_R = \sqrt{25 - 9}$$
$$E_R = \sqrt{16}$$
$$E_R = 4 \text{ volts}$$

TOTAL CURRENT IN A SERIES RESISTIVE-CAPACITIVE (RC) CIRCUIT (SEE FIG. 22)

Current in a series RC circuit is the same throughout the circuit. The same current flows through the resistor and the capacitor plates at the same amplitude.

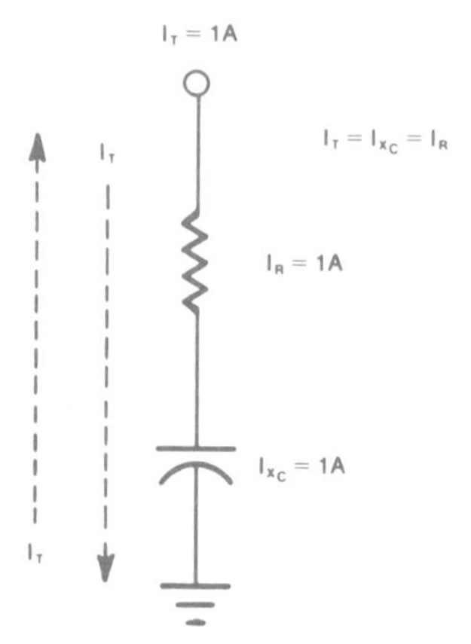

Fig. 22. Current Flow in a Series RC Circuit

IMPEDANCE IN A SERIES RESISTIVE-CAPACITIVE (RC) CIRCUIT (SEE FIG. 23)

In a series RC circuit, impedance is calculated using the Pythagorean theorem (vector sum). This theorem is stated electrically as $Z^2 = R^2 + X_C{}^2$ or $Z = \sqrt{R^2 + X_C{}^2}$. In Fig. 23, the horizontal side of the triangle is plotted as resistance (R). The vertical side, which is negative in direction, is plotted as capacitive reactance (X_C). The reason for this is that the X_C voltage lags current in a capacitive circuit. The angle, being negative, is drawn in a clockwise direction. The plotted angle shown in the figure is sometimes called a phasor. Calculation procedures are as follows:

(1) *IF Z IS UNKNOWN* — *THEN:* $Z^2 = X_C{}^2 + R^2$

GIVEN:

$X_C = 3$ ohms

$R = 4$ ohms

$Z^2 = 3^2 + 4^2$

$Z = \sqrt{9 + 16}$

$Z = \sqrt{25}$

$Z = 5$ ohms

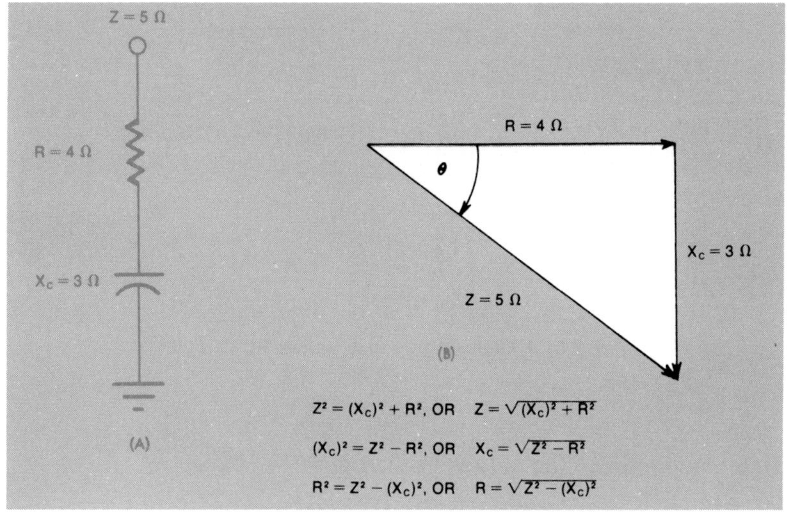

Fig. 23. Impedance in a Series RC Circuit

(2) *IF X_C IS UNKNOWN* *THEN:* $X_C{}^2 = Z^2 - R^2$

GIVEN:

Z = 5 ohms

R = 4 ohms

$$X_C{}^2 = 5^2 - 4^2$$
$$X_C = \sqrt{25 - 16}$$
$$X_C = \sqrt{9}$$
$$X_C = 3 \text{ ohms}$$

(3) *IF R IS UNKNOWN* *THEN:* $R^2 = Z^2 - X_C{}^2$

GIVEN:

Z = 5 ohms

X_C = 3 ohms

$$R^2 = 5^2 - 3^2$$
$$R = \sqrt{25 - 9}$$
$$R = \sqrt{16}$$
$$R = 4 \text{ ohms}$$

SERIES RC CIRCUIT PHASE ANGLES **(SEE FIG. 24)**

The phase angle in a series RC circuit is the relationship between applied or total voltage and current. The phase angle can be calculated using the impedance vector or the voltage vector. Since capacitive reactance X_C is plotted in the negative direction from horizontal, it takes a negative reference sign ($-X_C$). To calculate the phase angle using the impedance triangle the following formula is used:

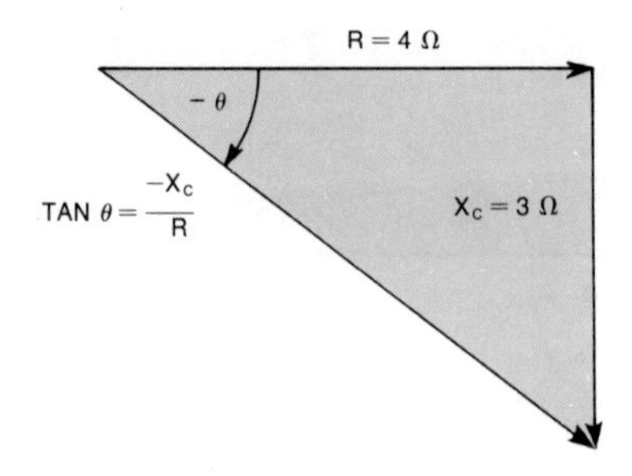

(A) IMPEDANCE TRIANGLE

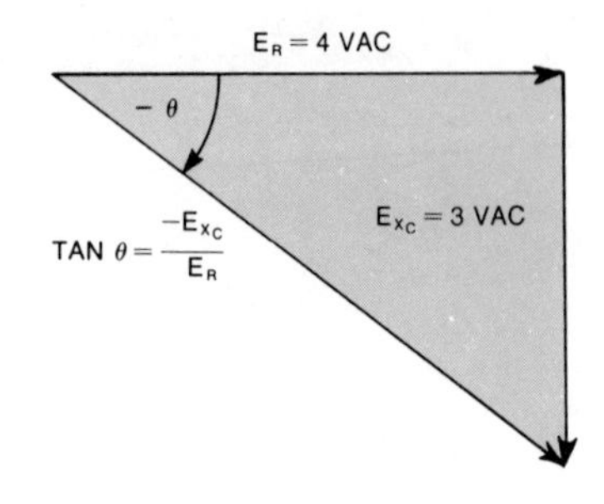

(B) VOLTAGE TRIANGLE

Fig. 24. Phase Angles in a Series RC Circuit

$$\text{TAN } \theta = \frac{-X_C}{}$$

Calculating, using the values in the figure, we have:

$$\text{TAN } \theta = \frac{-3}{4}$$

$$\text{TAN } \theta = -0.750$$

$$\theta = -36.9^\circ$$

Since voltage across the capacitor E_{X_C} is plotted in the negative direction from horizontal, it takes a negative reference sign or $(-E_{X_C})$. To calculate the phase angle using the voltage triangle, the following formula is used:

$$\text{TAN } \theta = \frac{-E_{X_C}}{E_R}$$

Calculating using the values in the figure, we have:

$$\text{TAN } \theta = \frac{-3}{4}$$

$$\text{TAN } \theta = -0.750$$

$\theta = -36.9^\circ$, the same phase angle as calculated using the impedance vector.

Note: *The phase angle can be calculated in a fraction of a minute with a slide rule of the decimal trig type. Having found this angle θ, it is then extremely quick and easy to find either impedance or total voltage of a series RC circuit because both are represented as the hypotenuse of a right triangle. Users of slide rules may prefer this method to using the root-mean-square equations because it avoids the use of squares and square roots. These methods are explained in detail in Appendix D: Solving Electrical Problems with Trigonometry Tables or the Slide Rule.*

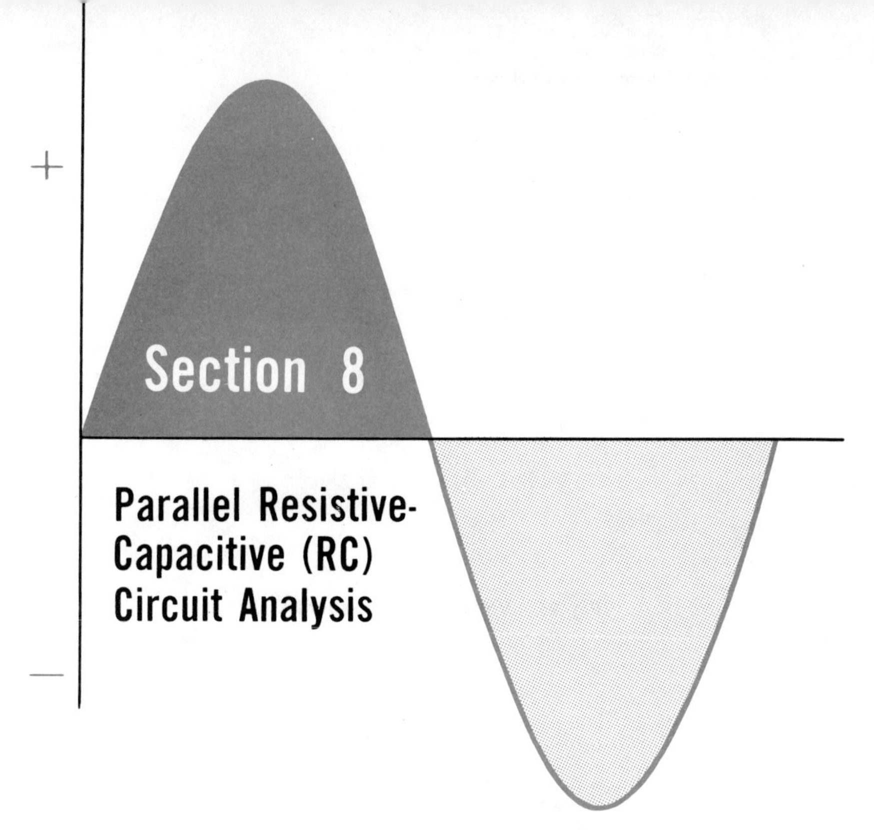

Parallel Resistive-Capacitive (RC) Circuit Analysis

In a parallel RC circuit the resistive and capacitive currents sum vectorially to provide total current. The phase angle created by this summation leads voltage. This reaction is useful in control devices and other circuits where time delays are required.

VOLTAGE IN A PARALLEL RESISTIVE-CAPACITIVE (RC) CIRCUIT (SEE FIG. 25)

Voltage in a parallel RC circuit is the same throughout each parallel leg. Applied voltage, voltage drop across the capacitor, and voltage drop across the resistor are of the same amplitude.
phase.

TOTAL CURRENT IN A PARALLEL RESISTIVE-CAPACITIVE (RC) CIRCUIT (SEE FIG. 26)

Although the voltage drop across the resistive leg and the capacitive leg of a parallel RC circuit is the same, current flow is different. The reason for this is that the capacitive current I_{X_C} leads capacitive voltage E_{X_C} by 90 degrees. When I_{X_C} is at maximum amplitude, E_{X_C}

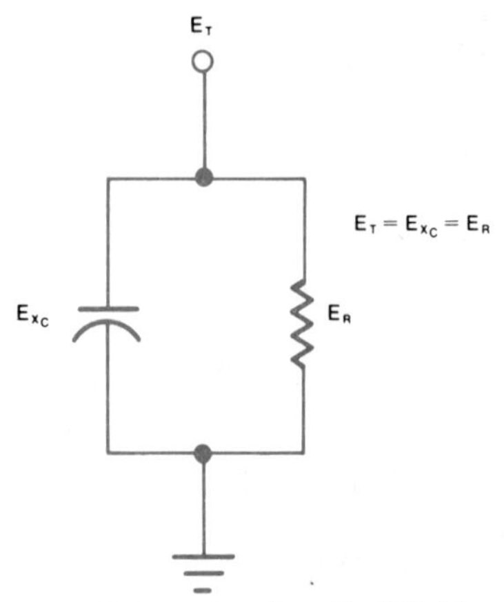

Fig. 25. Voltage in a Parallel RC Circuit

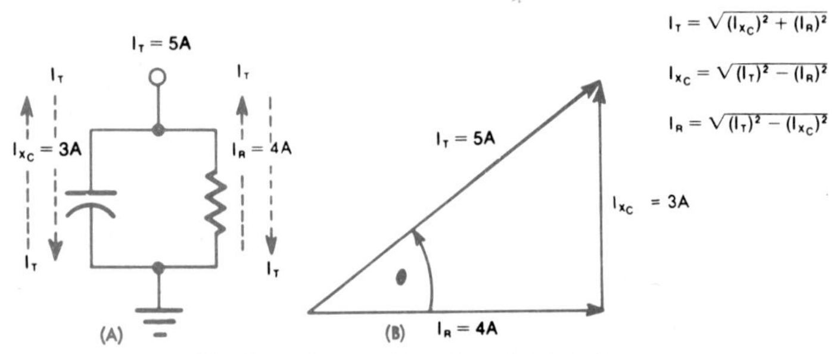

Fig. 26. Current in a Parallel RC Circuit

is at minimum amplitude. Resistive current I_R and voltage E_R are in phase. To find the total current in a resistive-capacitive (RC) circuit, the Pythagorean theorem (vector analysis) must be used. Electrically stated, the theorem is $I_T^2 = I_{X_C}^2 + I_R^2$, or $I_T = \sqrt{I_{X_C}^2 + I_R^2}$.

In Fig. 26 a current triangle is illustrated to depict current in an RC circuit. Note that I_{X_C} is plotted in the upward direction, because it leads E_{X_C}. Calculations, using the values in the figure, are as follows:

(1) *IF I_T IS UNKNOWN* — *THEN:* $I_T^2 = I_{X_C}^2 + I_R^2$

GIVEN:

$I_{X_C} = 3$ amps

$I_R = 4$ amps

$I_T^2 = 3^2 + 4^2$

$I_T = \sqrt{9 + 16}$

$I_T = \sqrt{25}$

$I_T = 5$ amps

(2) *IF I_{X_C} IS UNKNOWN* *THEN:* $I_{X_C}^2 = I_T^2 - I_R^2$

GIVEN:

$I_T = 5$ amps

$I_R = 4$ amps

$$I_{X_C}^2 = 5^2 - 4^2$$
$$I_{X_C} = \sqrt{25 - 16}$$
$$I_{X_C} = \sqrt{9}$$
$$I_{X_C} = 3 \text{ amps}$$

(3) *IF I_R IS UNKNOWN* *THEN:* $I_R^2 = I_T^2 - I_{X_C}^2$

GIVEN:

$I_T = 5$ amps

$I_{X_C} = 3$ amps

$$I_R^2 = 5^2 - 3^2$$
$$I_R = \sqrt{25 - 9}$$
$$I_R = \sqrt{16}$$
$$I_R = 4 \text{ amps}$$

IMPEDANCE IN A PARALLEL RESISTIVE-CAPACITIVE (RC) CIRCUIT (SEE FIG. 27)

Impedance (Z) in a parallel RC circuit equals the applied voltage E_T divided by the total current I_T. The following calculations are made using the values in Fig. 27.

$$Z = \frac{E_T}{I_T}$$

$$Z = \frac{5 \text{ volts}}{5 \text{ amps}}$$

$$Z = 1 \text{ ohm}$$

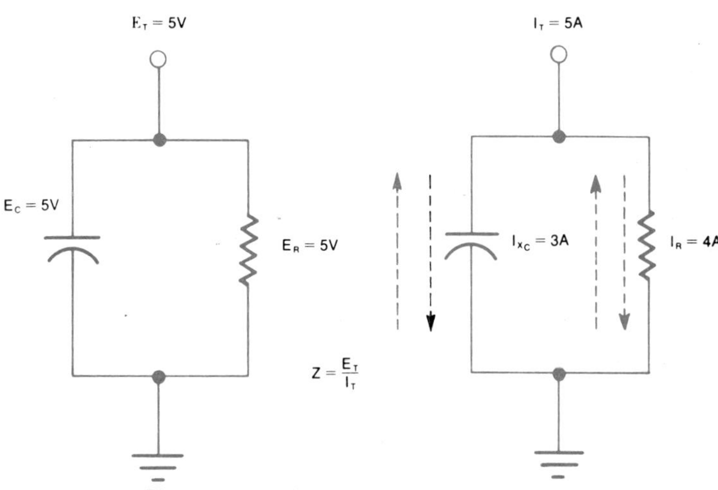

Fig. 27. Impedance in a Parallel RC Circuit

PARALLEL RC CIRCUIT PHASE ANGLES (SEE FIG. 28)

The phase angle in a parallel RC circuit is the relationship between the applied voltage E_T and reactive current I_{X_C}. Since the applied voltage E_T and resistive current I_R are in phase, the reactive current I_{X_C} and resistive current are utilized for phase angle calculations. In Fig. 28, the reactive current I_{X_C} is plotted upward because I_{X_C} leads E_T by 90 degrees. The sign it takes is positive $(+I_{X_C})$. The following calculations are made using the values in Fig. 28.

$$\text{TAN } \theta = \frac{I_{X_C}}{I_R}$$

$$\text{TAN } \theta = \frac{3}{4}$$

$$\text{TAN } \theta = 0.750$$

$$\theta = 36.9^\circ$$

Note: *The phase angle can be calculated in a fraction of a minute with a slide rule of the decimal trig type. Having found this angle θ, it is then extremely quick and easy to find the total current of a parallel RC circuit because it is represented as the hypotenuse of a right triangle. Users of slide rules may prefer this method to using the root-mean-square equation because it avoids the use of squares and square roots. These methods are explained in detail in Appendix D: Solving Electrical Problems with Trigonometry Tables or the Slide Rule.*

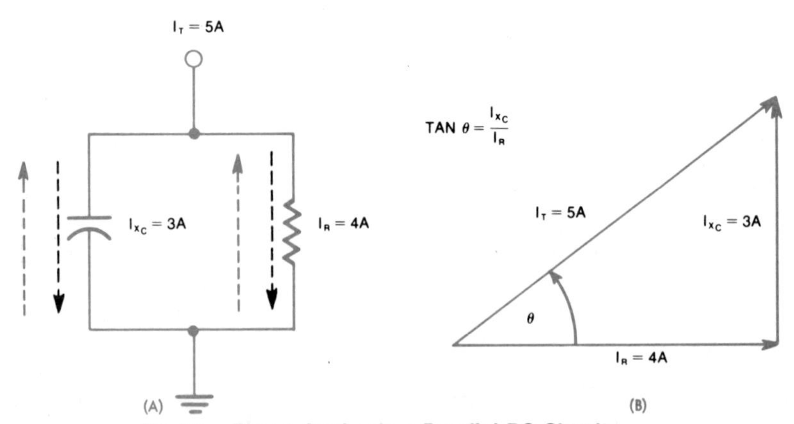

Fig. 28. Phase Angles in a Parallel RC Circuit

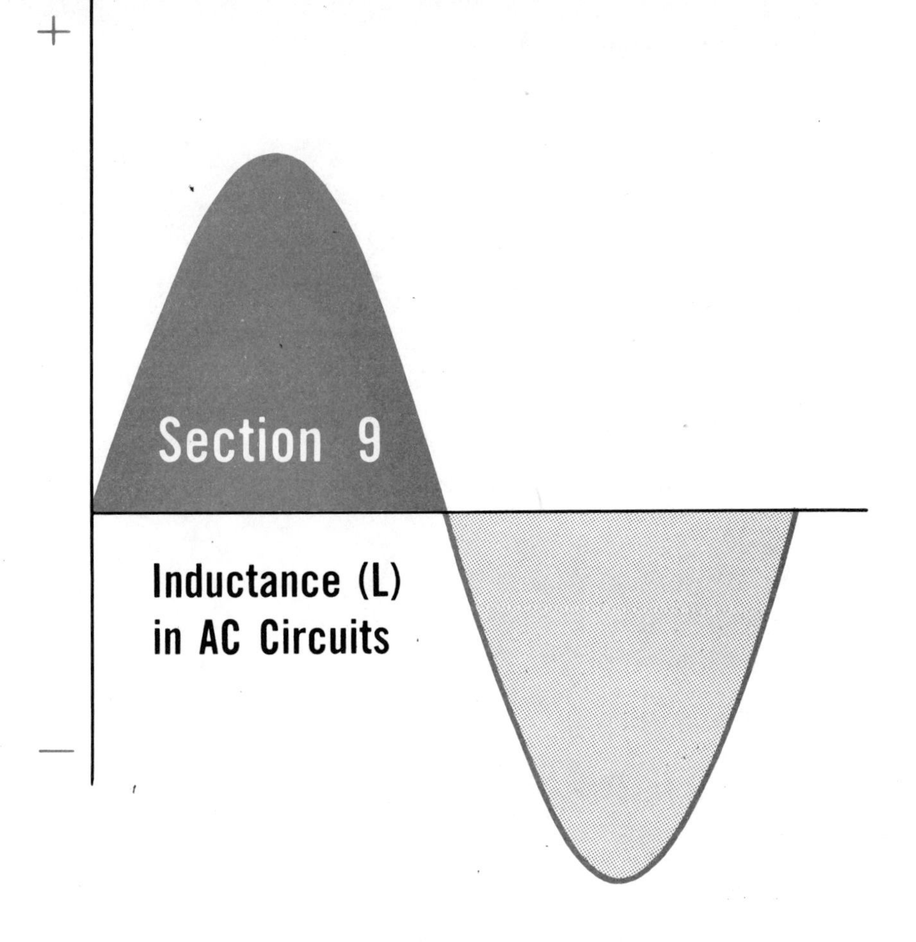

The inductor (like the resistor and the capacitor) is one of the basic components in an ac circuit. The inductor is also called a *reactor*. Its basic purpose is to react to a change in current flow. With this reaction to a change in current, the inductor is useful for sine wave propagation and signal development.

THE INDUCTOR (SEE FIG. 29)

An inductor is a coil of wire used to induce voltage at a known rate when current in the wire varies. Inductance is the ability of the inductor to provide this induced voltage. The inductor, like the capacitor, is called a *reactor* because it alternately stores energy and then delivers the energy back to the circuit. In the next several paragraphs we shall look at the inductor and determine what special attributes it has and what some of its problems are.

When current increases through a conductor, the magnetic field around the conductor expands. If the current is alternating and the

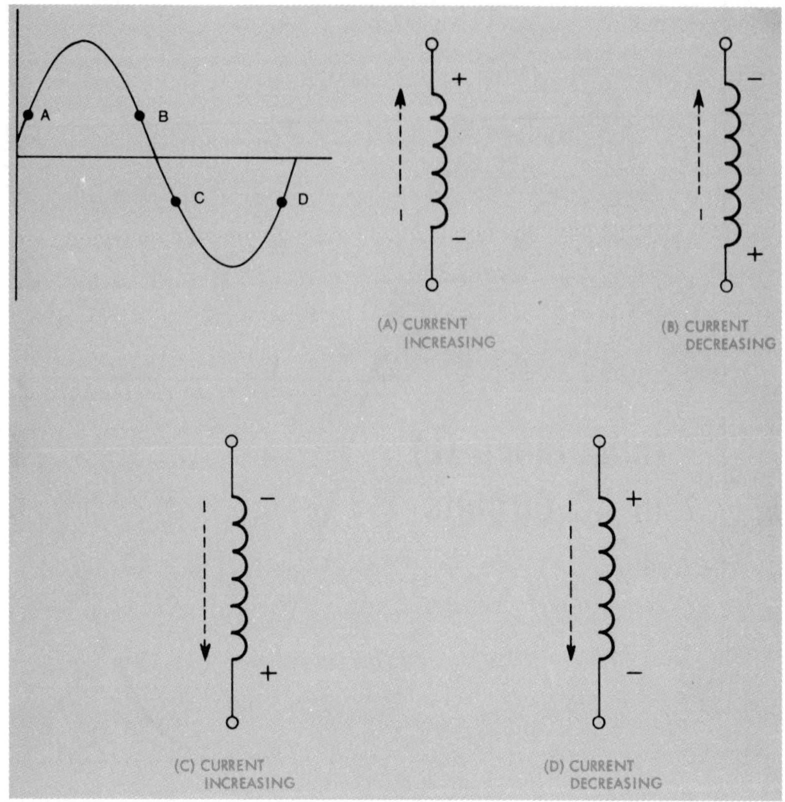

Fig. 29. Current and Voltage Polarity Relationship in a Coil (Lenz's Law)

wire is coiled, the current produces alternately expanding and collapsing magnetic fields around the coil. This field induces a voltage in the inductor that is opposite in direction to the applied voltage.

The self-induced voltage tends to keep the current moving in the same direction when the applied voltage decreases and opposes the current when the applied voltage increases. Opposition to current flow is called back- or counter-EMF.

The property of the coil which opposes a change in current is called inductance. Since the inductance is self-imposed this inductance is called *self-inductance*. The entire phenomenon is stated in Lenz's law: "When a voltage is induced in a coil as a result of a variation of the magnetic field with respect to the coil, the induced voltage is in such a direction as to oppose the current charge that caused the magnetic variation."

Fig. 29 illustrates the current and voltage polarity relationships in a coil with alternating current flowing. In Fig. 29(A), current flows in the bottom of the coil and the induced voltage opposes the increase. In Fig. 29(B), current decreases and the induced voltage polarity in the coil changes to oppose the decrease. In Fig. 29(C), current flows in the top of the coil and the induced voltage opposes the increase. In Fig. 29(D), current decreases and the induced voltage polarity of the coil changes to oppose the decrease.

MUTUAL INDUCTANCE (SEE FIG. 30)

When two coils are placed side by side so that magnetic lines of force of one coil cut the turns of the other coil, the two coils are said to have mutual inductance.

In Fig. 30, the two coils are magnetically tied together with magnetic flux lines. As current varies in the primary coil (L_1), voltage is induced in the secondary coil (L_2). This action is typical of transformer operation.

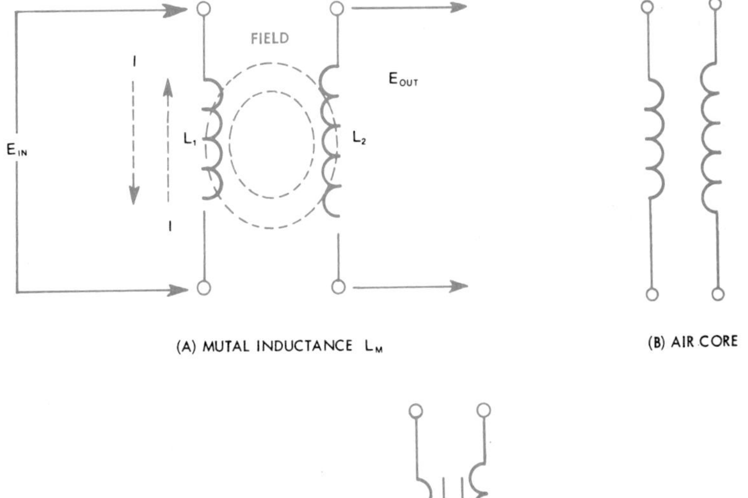

(A) MUTAL INDUCTANCE L_M

(B) AIR CORE

(C) IRON CORE

Fig. 30. Mutual Inductance of Two Coils

FACTORS AFFECTING INDUCTANCE

Several factors affect inductance. The first of these factors is the *diameter to length ratio of the coil.* If everything else remains constant, the inductance of the coil varies directly with an increase in the diameter of the coil and inversely with an increase in the length of the coil.

The second thing that affects coil inductance is the *number of turns in the coil.* The greater the number of turns in the coil, the stronger the magnetic field is and the greater the inductance. Inductance is directly proportional to the square of the number of turns in the coil.

The third thing that affects the inductance of a coil is the *type of core* within the coil. Inductance is directly proportional to the *core permeability*. Coils that have non-magnetic cores have less inductance than magnetic core coils, with all other inductance factors being equal.

The final thing that affects coil inductance is the *method of winding*. Inductance can be varied by changing the spaces between the windings. Coils can be wound in layers, have spaces between windings or wound close together. Inductance is greatest when the windings are close and numerous.

FLUX LINKAGE (SEE FIG. 31)

Flux linkage is the sum of all the lines of force (flux lines) that surround a conductor, whether it is a single wire or a coil. In Fig.

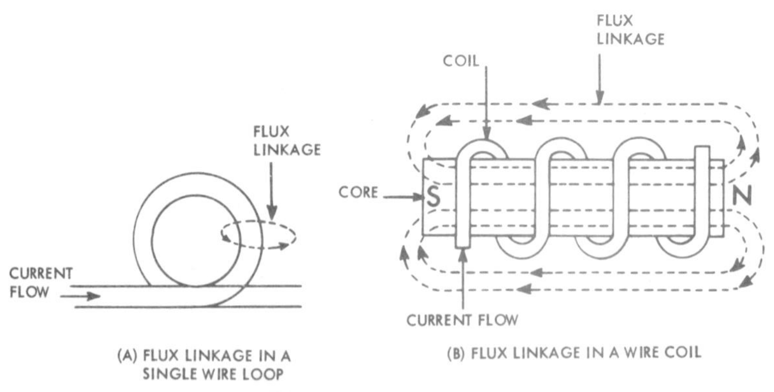

Fig. 31. Flux Linkages

31(A), the flux linkage encircles the single wire. In Fig. 31(B), the flux lines encircle all the loops of the coil. Flux linkage is the product of the number of flux lines times the number of turns of the coil. Inductance may also be said to be the number or flux linkages per unit of current.

INDUCTOR TYPES (SEE FIG. 32)

Inductors are manufactured in many shapes and forms. The basic form is a coil of wire with no core (air core). Other types have iron cores and ferrite cores. The core is very important because that is where the magnetic lines of force are the most dense. The shape of an inductor is dependent on the size and packaging required for the inductor to fit in an allotted space.

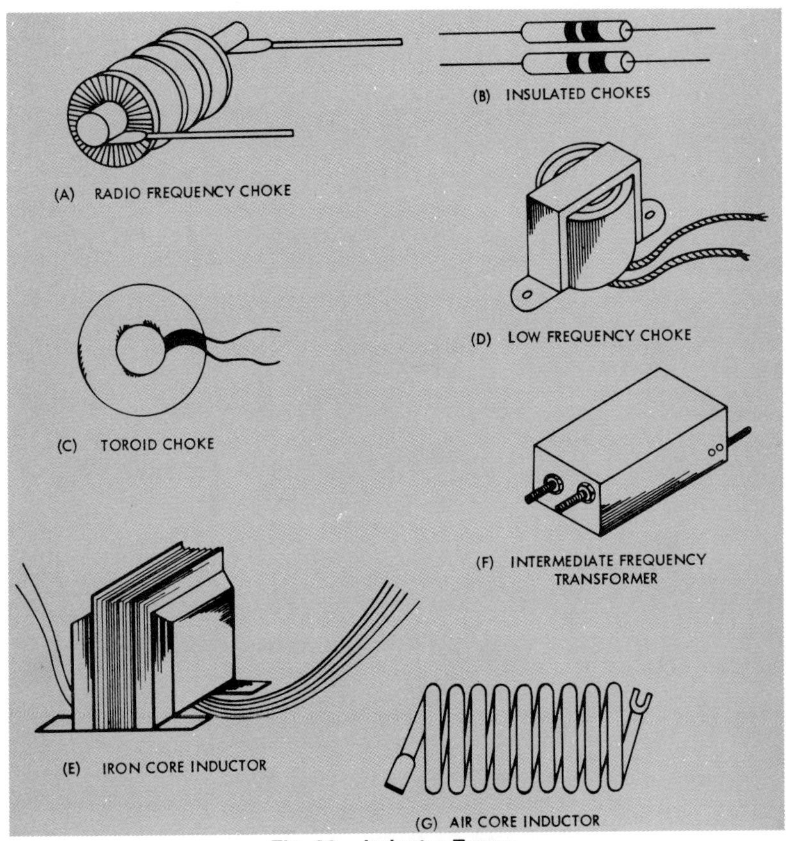

Fig. 32. Inductor Types

Types of inductors are named by the job they do, such as intermediate and interstage transformers, audio and radio frequency transformers (chokes), and power transformers. Inductors are often called *transformers, chokes,* or *reactors.*

If an inductor is needed to change variable inductance it may be accomplished in one of two ways. Varying permeability of an inductor may be achieved by attaching a cylindrical metal slug which may be screwed into or out of the coil core. The second method, varying with the number of turns, may be achieved by selecting the number of turns on the coil in a manner similar to that of a rheostat. A tap is then connected to the selected point.

INDUCTOR SYMBOLS (SEE FIG. 33)

Inductor symbols vary with manufacturer. However, the stand-

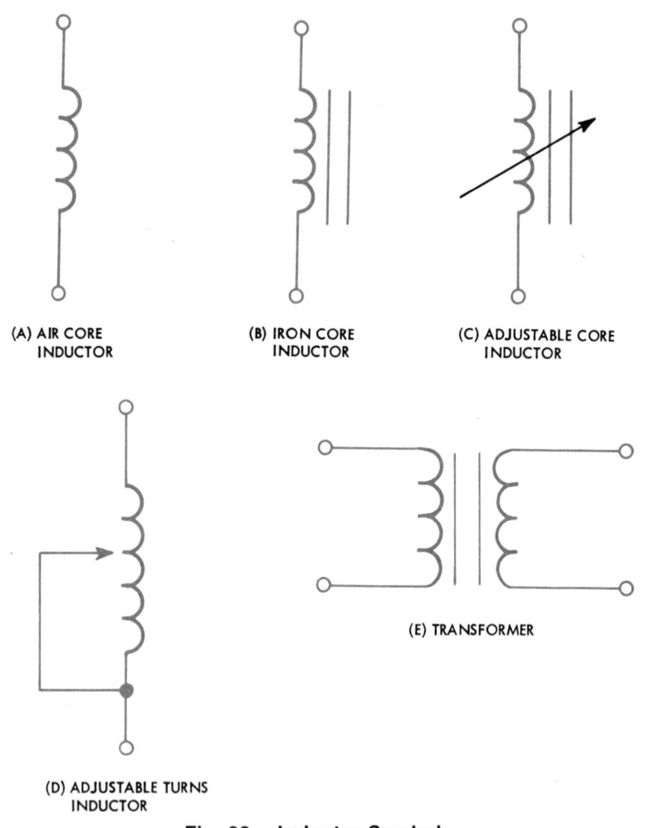

Fig. 33. Inductor Symbols

ard symbols according to military specifications are shown in Fig. 33. These are also recognized throughout industry.

QUALITY (Q) OF A COIL

The quality (Q) of a coil is the coil's ability to produce self-induced voltage. The internal resistance of the coil reduces the current and therefore its voltage-producing ability.

At low frequencies the resistance is merely the dc resistance of the wire. At higher frequencies, the inductive reactance of the coil increases ($X_L = 2\pi fL$). Since the Q of a coil is the ratio of reactive power to real power, $\left(Q = \frac{P_{reactive}}{P_{real}}\right)$, the formula for the Q of a coil can be reduced as follows:

$$Q = \frac{P_{reactive}}{P_{real}}$$

$$\text{or } Q = \frac{I^2X_L}{I^2R}, \text{ or } \frac{X_L}{R}$$

As an example, a coil with an inductive reactance of 200 ohms and a dc resistance of 10 ohms would have a Q of 20 (this value is numerical only and reflects the ratio $\left.\frac{X_L}{R}\right)$.

$$Q = \frac{X_L}{R}$$

$$Q = \frac{200}{10}$$

$$Q = 20$$

INDUCTORS IN SERIES AC CIRCUITS (SEE FIG. 34)

Inductors in series ac circuits are added in the same manner as resistors in series. Solution is as follows:

GIVEN:

NOTE: mH = millihenry

$L_1 = 10$ mH

$L_2 = 20$ mH

$L_3 = 20$ mH

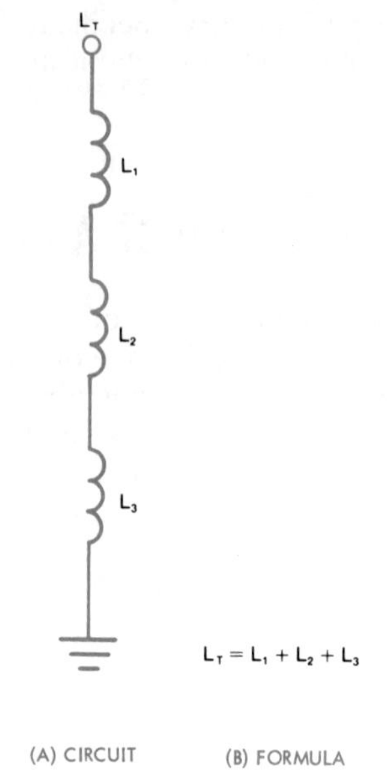

Fig. 34. Inductors in Series AC Circuits

FIND: L_T

$$L_T = L_1 + L_2 + L_3$$

$$L_T = 10 + 20 + 20$$

$$L_T = 50 \text{ mH}$$

INDUCTORS IN PARALLEL AC CIRCUITS (SEE FIG. 35)

Inductors in parallel ac circuits are added using the reciprocal formula, in the same manner as resistors in parallel. In Fig. 35 note that there are two formulas. The first provides the reciprocal formula to be used for more than two inductors. Solution is as follows:

GIVEN: *NOTE:* mH = millihenry

$L_1 = 10$ mH

$L_2 = 20$ mH

$L_3 = 20$ mH

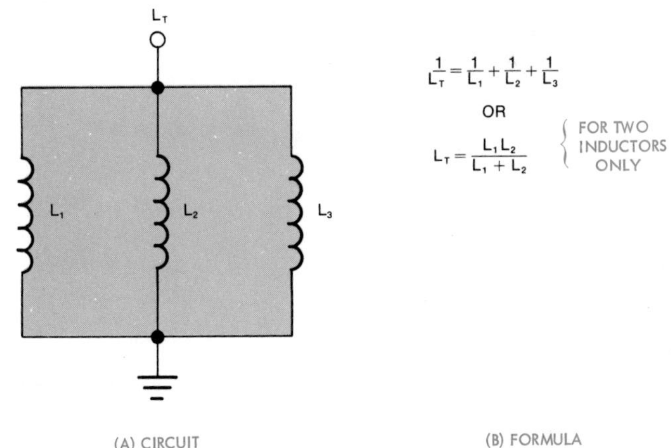

Fig. 35. Inductors in Parallel AC Circuits

FIND: L_T

$$\frac{1}{L_T} = \frac{1}{L_1} + \frac{1}{L_2} + \frac{1}{L_3}$$

$$\frac{1}{L_T} = \frac{1}{10} + \frac{1}{20} + \frac{1}{20}$$

$$\frac{1}{L_T} = \frac{2}{20} + \frac{1}{20} + \frac{1}{20} = \frac{4}{20}$$

$$L_T = \frac{20}{4} = 5 \text{ mH}$$

Calculations for total inductance in parallel ac circuits with only two inductors is as follows:

GIVEN: *NOTE:* mH = millihenry

$$L_1 = 10 \text{ mH}$$

$$L_2 = 20 \text{ mH}$$

FIND: L_T

$$L_T = \frac{L_1 L_2}{L_1 + L_2}$$

$$L_T = \frac{10 \times 20}{10 + 20}$$

$$L_T = \frac{200}{30}$$

$$L_T = 6.67 \text{ mH}$$

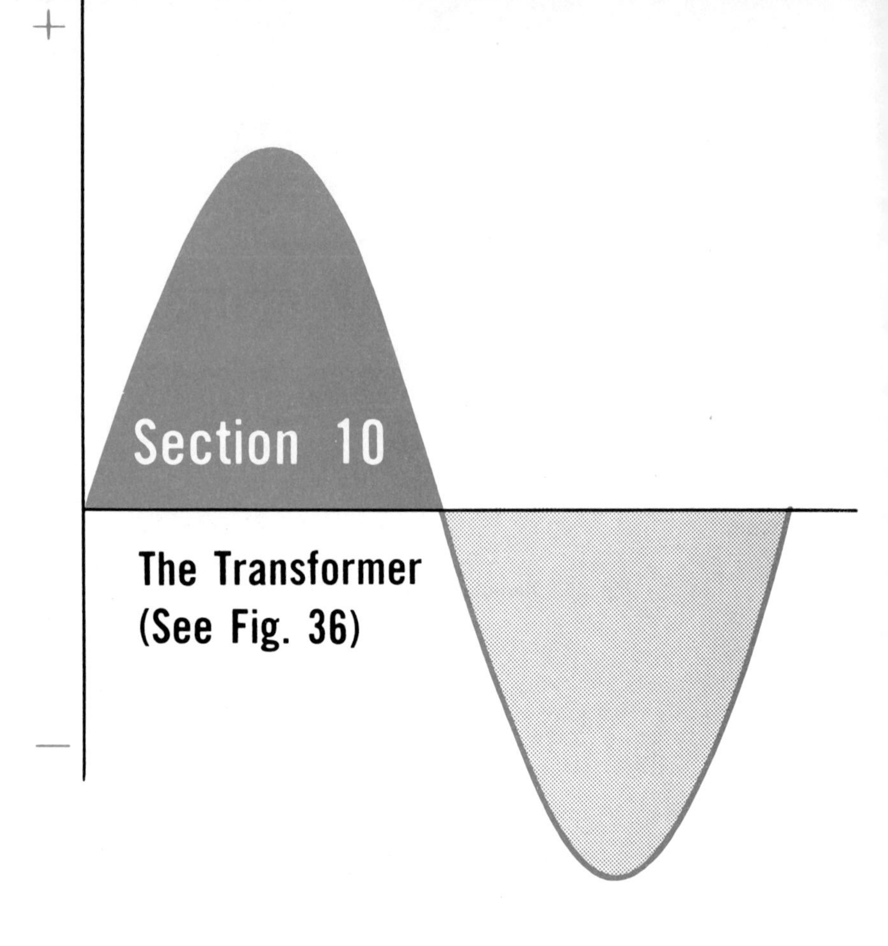

Section 10

The Transformer (See Fig. 36)

The transformer is a device that transfers power from one coil of wire to another. It does so by mutual inductance. Mutual inductance is the phenomenon that allows flux lines of an inductor to induce voltage in another inductor close to it when current changes.

A transformer is usually constructed to have a primary coil and one or several secondary coils. AC power is applied to the primary of the transformer and induced into the secondary, where it is used to produce current through a load.

Power in the primary of a transformer is equal to power out of the secondary of the transformer with accepted losses.

One of these losses is *eddy currents.* Eddy currents are circular currents in an iron core transformer that are induced in the transformer because its core is made of a conductive material (iron). The eddy currents represent a power loss which increases with frequency.

Hysteresis is another power loss and is due to the additional power required to reverse the magnetic field as the alternations take place.

Other cores are powdered iron and laminated iron. These core

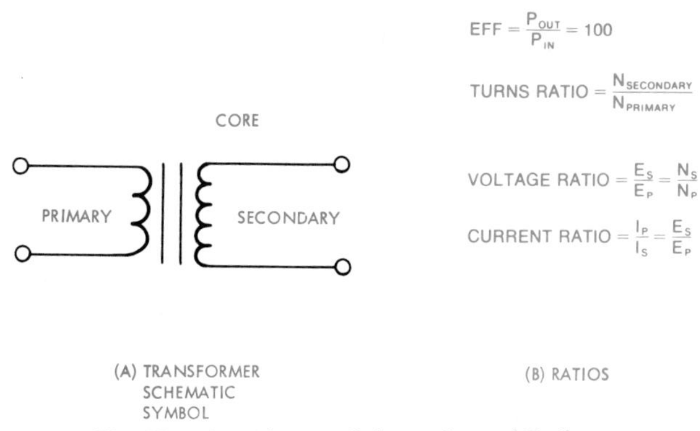

Fig. 36. Transformer Schematic and Ratios

types help to eliminate eddy currents. A *variac* is a transformer with variable inductance. It is wired so that the output can vary from 0 to a specified number of volts. Its primary and secondary coils are wound from a single wire. This variable type transformer is also called an *autotransformer*.

TRANSFORMER CALCULATIONS

Efficiency in a transformer is calculated by the ratio of power out to power in. That is,

$$\text{EFFICIENCY} = \frac{P_{OUT}}{P_{IN}} \times 100 \qquad \text{(Expressed as a percentage)}$$

A perfect transformer would have the same power out as in. This, of course, is never the case, so efficiency is under this value.

Other ratios which are peculiar to transformers are turns ratio, voltage ratio, and current ratio. Turns ratio has to do with the number of turns in the primary and secondary coils. That is,

$$\text{TURNS RATIO} = \frac{N_{SECONDARY}}{N_{PRIMARY}}$$

$$\text{EXAMPLES: } \frac{10}{1} = 10{:}1 \qquad \frac{1}{10} = 1{:}10$$

$$\frac{2}{2} = 1{:}1$$

A transformer whose secondary coil has a larger number of turns than the primary coil is called a *step-up transformer*. A transformer whose secondary coil has fewer number of turns than the primary coil is called a *step-down transformer.*

The voltage ratio is directly proportional to the turn ratio. That is,

$$\frac{E_{SECONDARY}}{E_{PRIMARY}} = \frac{N_{SECONDARY}}{N_{PRIMARY}}$$

$$\text{or } \frac{E_S}{E_P} = \frac{N_S}{N_P}$$

EXAMPLE:

$$\frac{10 \text{ volts}}{5 \text{ volts}} = \frac{100 \text{ turns}}{50 \text{ turns}}$$

Current ratio is inversely proportional to the voltage ratio. That is,

$$\frac{I_{PRIMARY}}{I_{SECONDARY}} = \frac{E_{SECONDARY}}{E_{PRIMARY}}, \quad \text{or } \frac{I_P}{I_S} = \frac{E_S}{E_P}$$

EXAMPLE:

$$\frac{2 \text{ amps}}{1 \text{ amps}} = \frac{10 \text{ volts}}{5 \text{ volts}}$$

Note that power in this ratio is equal primary and secondary. ($I_P E_P = I_S E_S$)

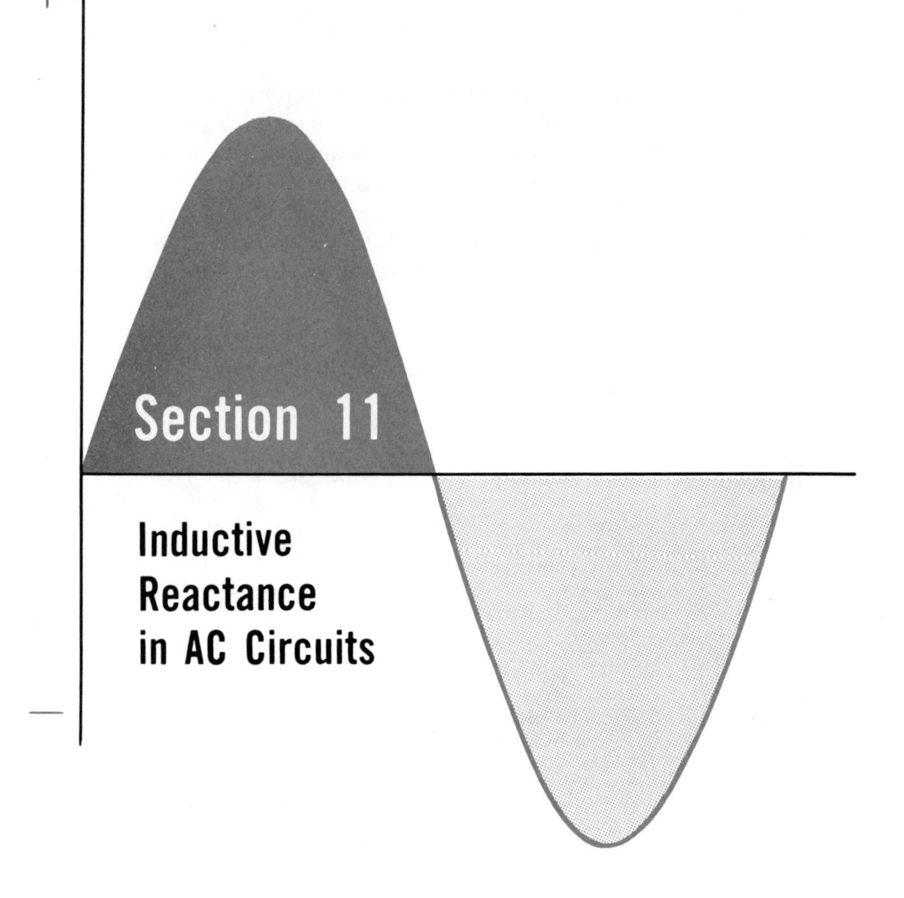

Section 11

Inductive Reactance in AC Circuits

Inductive reactance is opposition to current flow in an inductive ac circuit. It is calculated in ohms the same as resistance; however, it cannot be measured as resistance with an ohmmeter (only calculated). The reason for this is that an inductor responds to current changing with time. Inductive opposition to current flow is a ratio of voltage to current in an ac circuit.

Inductive reactance is distinguished from pure resistance by the symbol X_L. Inductive reactance is affected by two variables; inductance of the coil (in henrys) and the frequency of the ac current flowing in the circuit (in hertz). For pure sinusoidal circuits, this opposition to current flow (inductive reactance) may be calculated with the formula:

$$X_L = 2\pi fL$$

WHERE: X_L = Inductive reactance
π = 3.1416 (constant)
f = Frequency of the current in hertz
L = Inductance of the coil in henrys

It may be noted that an increase of either frequency or inductance will increase inductive reactance. Likewise, a decrease of frequency or inductance will decrease inductive reactance. This increase or decrease is linear. Low frequency ac currents applied to an inductor will create low reactance. High frequency ac applied to an inductor will create infinite reactance and will react as an open circuit. The same is true for an increase or decrease in inductance. Fig. 37 illustrates inductive reactance in relation to a change in frequency or inductance.

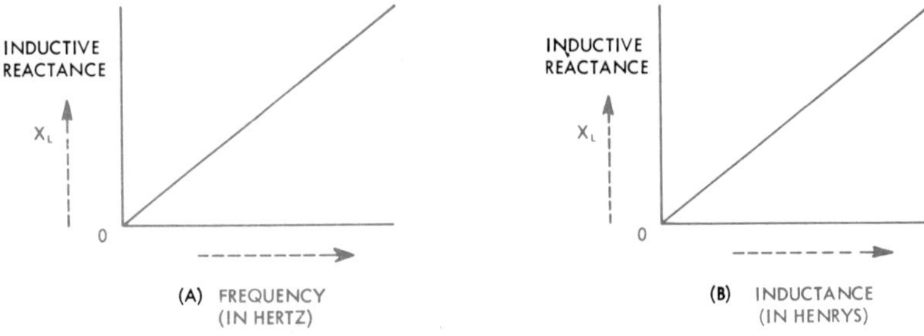

Fig. 37. Change in Inductive Reactance as Frequency or Inductance Vary

PURE INDUCTIVE PHASE ANGLES (SEE FIG. 38)

In a pure inductive ac circuit, the voltage sine wave leads the current sine wave by 90 degrees at any instantaneous point in time. It could also be said that under these circumstances, current lags voltage. This is explained simply: When the voltage is at 90 degrees in time, current is at 0 degrees in time. When voltage is at 180 degrees in time, current is at 90 degrees in time etc.

At the 0 and 180 degree position in time, voltage changes direction. At the 90 and 270 degree position in time, current changes direction. However, the directions are opposite. As the voltage increases in amplitude, current is decreasing in amplitude, and vice-versa. With pure inductance this is true; however, inserting resistance in the circuit causes the phase relationship of voltage and current to vary. This change is explained under RL series and parallel circuit analysis.

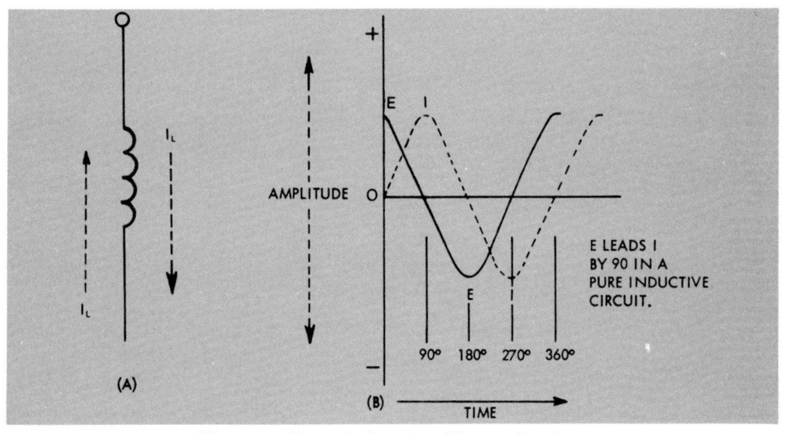

Fig. 38. Pure Inductive Phase Angles

RL TIME CONSTANT (SEE FIG. 39)

The time constant in a resistive-inductive circuit is actually the time that it takes for the current to change by 63.2%. An RL circuit contains at least one inductor and one resistor. Fig. 39 illustrates this circuit, which has 1.0 ampere of steady-state current flow.

When the switch is closed, current starts to flow, and in one time constant increases from 0 to 0.632 ampere. In the second time constant, current increases by 63.2% of the remaining steady state current as shown below.

STEADY STATE CURRENT	1.000
1st TIME CONSTANT	− 0.632
REMAINING	0.368 ampere

63.2% of 0.368 = 0.232 ampere increase in 2nd time constant

Current continues to increase for the 3rd, 4th, and 5th time constant in the same manner. After the 5th time constant, for practical application, current has reached the 1.0 ampere steady-state current.

While current is building up from 0 to 1.0, the inductor is opposing the change in current. The reaction of the RL circuit during this time is called *transient response*. The time constant (TC) in seconds is a ratio of inductance and resistance as shown below.

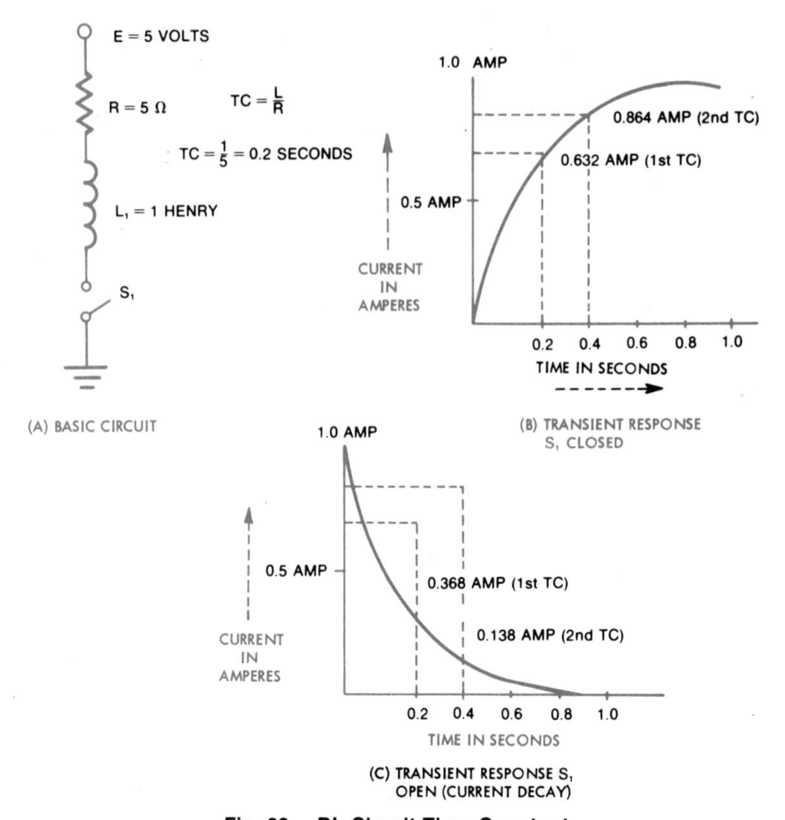

Fig. 39. RL Circuit Time Constant

$$TC = \frac{L}{R}$$

$$TC = \frac{1 \text{ henry}}{5 \text{ ohms}}$$

$$TC = 0.2 \text{ seconds}$$

If the closed circuit shown in Fig. 39(A) is now opened, the reaction of the circuit will be that the current will decay 63.2% in the first time constant, then by 63.2% of the remaining current in the second time constant.

The current will continue to decay in the same manner, and after the fifth time constant will, practically speaking, be zero.

The time constant grows larger with an increase in inductance and smaller with an increase in resistance.

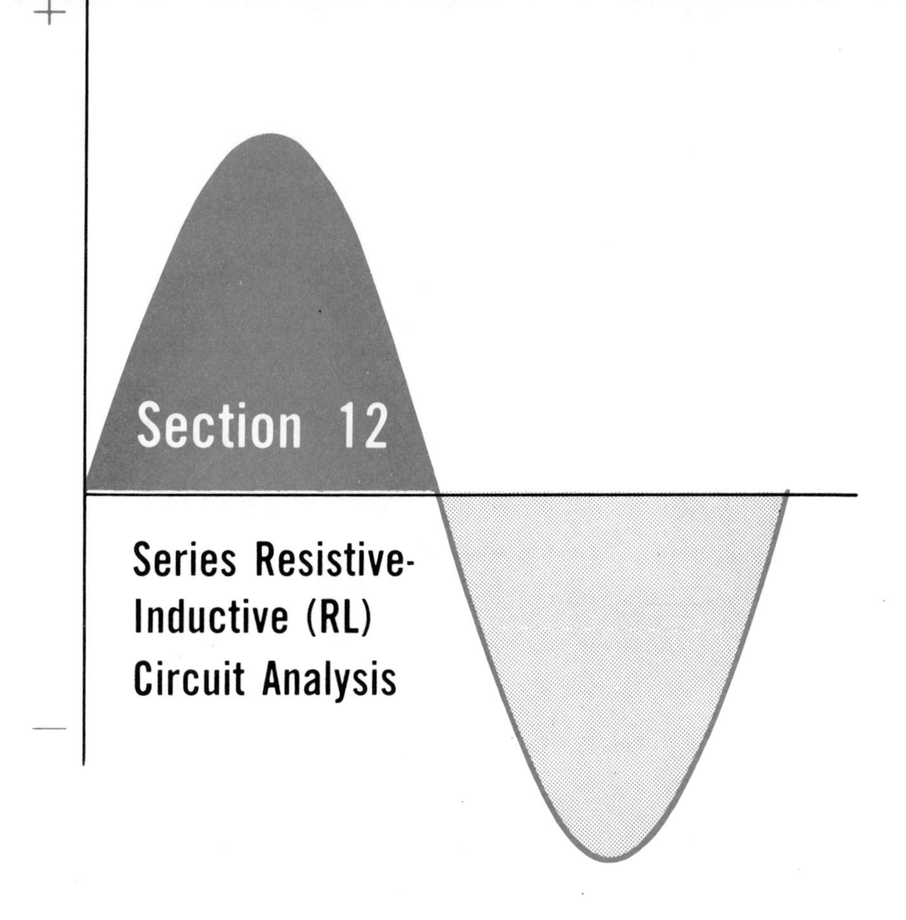

Section 12

Series Resistive-Inductive (RL) Circuit Analysis

Series RL circuits are similar in operation to series RC circuits. However, the phase angle created by the inductor's reactance to alternating current causes voltage to lead current. This reaction makes the RL circuit useful in circuits such as audio amplifiers for selective filtering, and for tuning.

TOTAL VOLTAGE IN A SERIES RESISTIVE-INDUCTIVE (RL) CIRCUIT (SEE FIG. 40)

In a series RL circuit, total voltage is calculated using the Pythagorean theorem (vector sum). This theorem is stated electrically as $E_T^2 = E_{X_L}^2 + E_R^2$, or $E_T = \sqrt{E_{X_L}^2 + E_R^2}$.

In Fig. 40, the horizontal side of the triangle is plotted as the voltage drop across the resistance (E_R). The vertical side, which is positive in direction, is plotted as the voltage drop across the inductor (E_{X_L}). The reason for this is that the X_L voltage leads current in an inductive circuit.

The angle, being positive, is drawn in a counterclockwise direc-

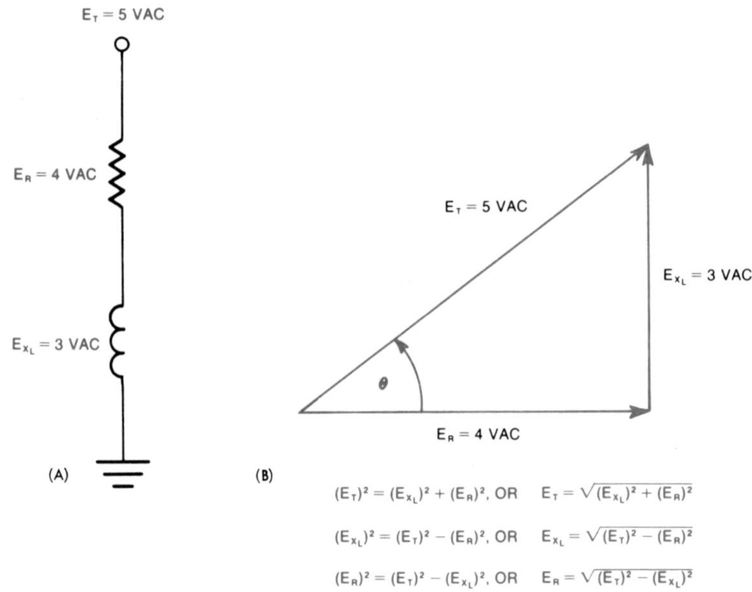

Fig. 40. Total Voltage in a Series RL Circuit

tion. The plotted angle shown in the figure is sometimes called a phasor. Calculations are as follows:

① *IF E_T IS UNKNOWN* *THEN:* $E_T^2 = E_{X_L}^2 + E_R^2$

GIVEN:

$E_{X_L} = 3$ volts

$E_R = 4$ volts

$$E_T^2 = 3^2 + 4^2$$
$$E_T = \sqrt{9 + 16}$$
$$E_T = \sqrt{25}$$
$$E_T = 5 \text{ volts}$$

② *IF E_{X_L} IS UNKNOWN* *THEN:* $E_{X_L}^2 = E_T^2 - E_R^2$

GIVEN:

$E_T = 5$ volts

$E_R = 4$ volts

$$E_{X_L}^2 = 5^2 - 4^2$$
$$E_{X_L} = \sqrt{25 - 16}$$
$$E_{X_L} = \sqrt{9}$$
$$E_{X_L} = 3 \text{ volts}$$

③ *IF E_R IS UNKNOWN* *THEN:* $E_R^2 = E_T^2 - E_{X_L}^2$

GIVEN:

$E_T = 5$ volts

$$E_R^2 = 5^2 - 3^2$$
$$E_R = \sqrt{25 - 9}$$

$E_{X_L} = 3$ volts

$E_R = \sqrt{16}$

$E_R = 4$ volts

CURRENT IN A SERIES RESISTIVE-INDUCTIVE (RL) CIRCUIT (SEE FIG. 41)

Current in a series R_L circuit is the same throughout the circuit. The same current flows through the resistor and the inductor at the same amplitude.

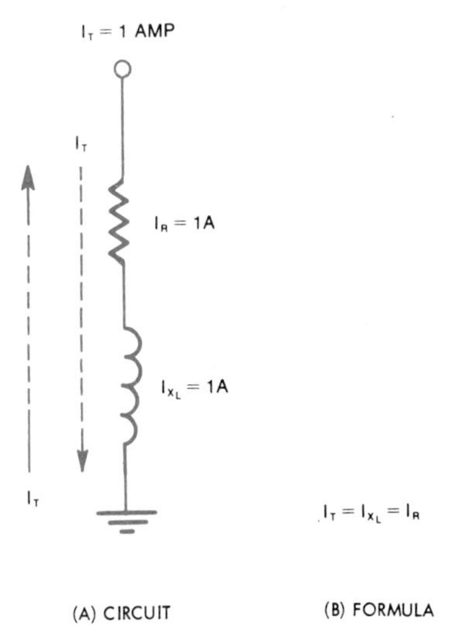

Fig. 41. Current Flow in a Series RL Circuit

IMPEDANCE IN A SERIES RESISTIVE-INDUCTIVE (RL) CIRCUIT (SEE FIG. 42)

In a series RL circuit, impedance is calculated using the Pythagorean theorem (vector sum). This theorem is stated electrically as $Z^2 = R^2 + X_L^2$, or $Z = \sqrt{R^2 + X_L^2}$. In Fig. 42, the horizontal side of the triangle is plotted as resistance (R). The vertical side, which is positive in direction, is plotted as inductive reactance (X_L). The reason for this is that the X_L voltage leads current in an inductive circuit.

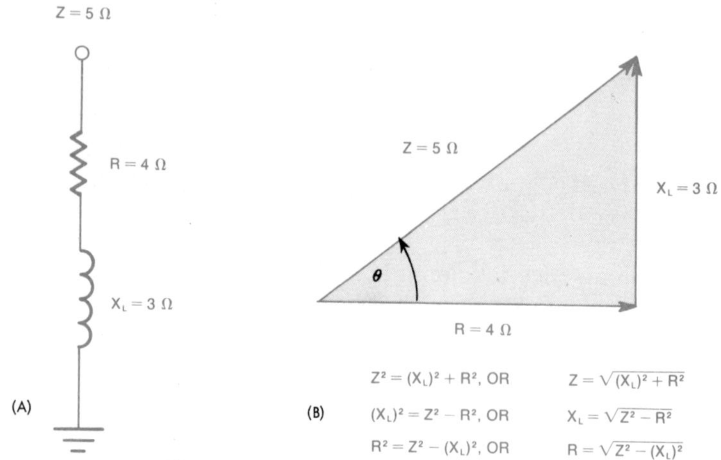

Fig. 42. Impedance in a Series RL Circuit

The angle, being positive, is drawn in a counterclockwise direction. The plotted angle shown in the figure is sometimes called a phasor. Calculation procedures are as follows:

(1) *IF Z IS UNKNOWN* *THEN:* $Z^2 = X_L{}^2 + R^2$

GIVEN:

$X_L = 3$ ohms

$R = 4$ ohms

$$Z^2 = 3^2 + 4^2$$
$$Z = \sqrt{9 + 16}$$
$$Z = \sqrt{25}$$
$$Z = 5 \text{ ohms}$$

(2) *IF X_L IS UNKNOWN* *THEN:* $X_L{}^2 = Z^2 - R^2$

GIVEN:

$Z = 5$ ohms

$R = 4$ ohms

$$X_L{}^2 = 5^2 - 4^2$$
$$X_L = \sqrt{25 - 16}$$
$$X_L = \sqrt{9}$$
$$X_L = 3 \text{ ohms}$$

(3) *IF R IS UNKNOWN* *THEN:* $R^2 = Z^2 - X_L{}^2$

GIVEN:

$Z = 5$ ohms

$X_L = 3$ ohms

$$R^2 = 5^2 - 3^2$$
$$R = \sqrt{25 - 9}$$
$$R = \sqrt{16}$$
$$R = 4 \text{ ohms}$$

SERIES RL CIRCUIT PHASE ANGLES (SEE FIG. 43)

The phase angle in a series RL circuit is the relationship between the applied or total voltage and the current.

The phase angle can be calculated using the impedance vector or the voltage vector. Since X_L is plotted in the positive direction from horizontal, it takes a + reference sign ($+X_L$). To calculate the phase angle using the impedance triangle the following formula is used:

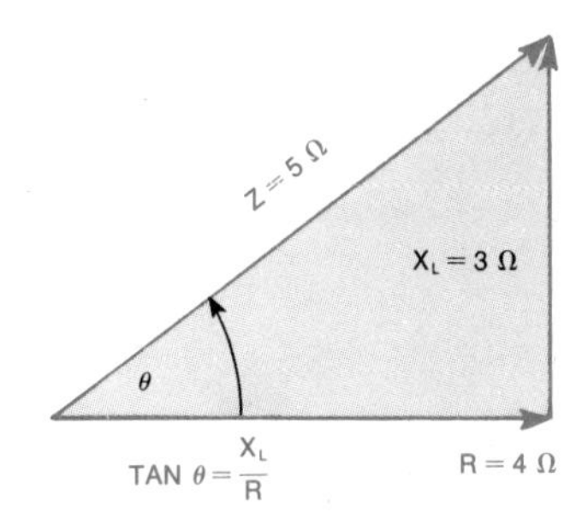

(A) IMPEDANCE TRIANGLE

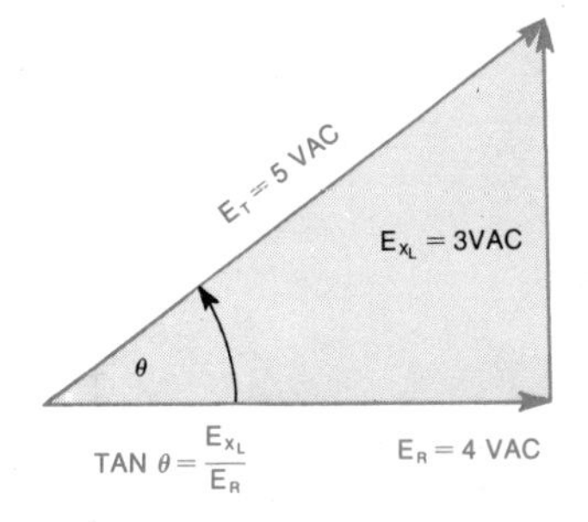

(B) VOLTAGE TRIANGLE

Fig. 43. Phase Angles in a Series RL Circuit

$$\text{TAN } \theta = \frac{X_L}{R}$$

Calculating using the values in Fig. 43, we have:

$$\text{TAN } \theta = \frac{3}{4}$$

$$\text{TAN } \theta = 0.750$$

$$\theta = 36.9°$$

Since E_{X_L} is plotted in the positive direction from horizontal, it takes a positive reference sign or ($+E_{X_L}$). To calculate the phase angle using the voltage triangle, the following formula is used:

$$\text{TAN } \theta = \frac{E_{X_L}}{E_R}$$

Calculating using the values in the figure 43, we have:

$$\text{TAN } \theta = \frac{3}{4}$$

$$\text{TAN } \theta = 0.750$$

$$\theta = 36.9^\circ$$

Note: *The phase angle can be calculated in a fraction of a minute with a slide rule of the decimal trig type. Having found this angle θ, it is then extremely quick and easy to find the total voltage or impedance of a series RL circuit because both are represented as the hypotenuse of a right triangle. These methods are explained in detail in Appendix D: Solving Electrical Problems with Trigonometry Tables or the Slide Rule. Users of slide rules may prefer the "trig" solution to using the root-mean-square equation because it avoids the use of squares and square roots.*

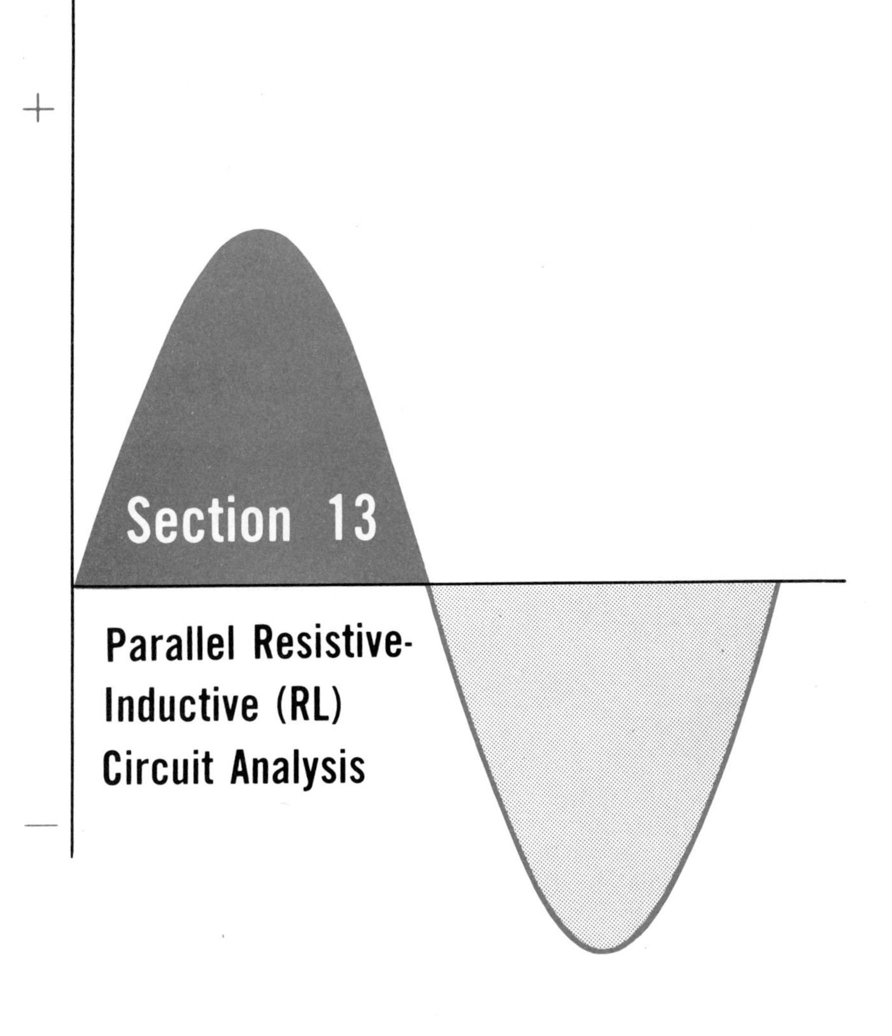

The parallel RL circuit is similar in operation to the parallel RC circuit. However, the phase angle created by the inductor's reactance to ac is summed vectorally, with the current lagging the voltage. These reactions are useful in audio and radio circuits.

VOLTAGE IN A PARALLEL RESISTIVE-INDUCTIVE (RL) CIRCUIT (SEE FIG. 44)

Voltage in a parallel RL circuit is the same throughout each parallel leg. Applied voltage, voltage drop across the inductor, and voltage drop across the resistor are of the same amplitude.

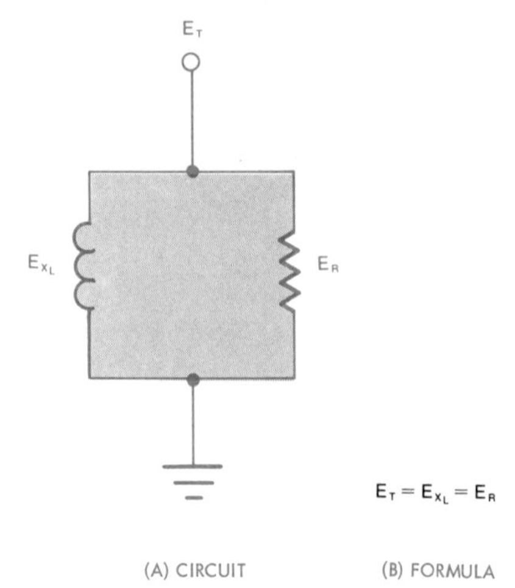

Fig. 44. Voltage in a Parallel RL Circuit

TOTAL CURRENT IN A PARALLEL RESISTIVE-INDUCTIVE (RL) CIRCUIT (SEE FIG. 45)

Although the voltage drop across the resistive leg and the inductive leg of a parallel RL circuit is the same, current flow is different. The reason for this is that the inductive current I_{X_L} lags inductive voltage E_{X_L} by 90 degrees.

When I_{X_L} is at a minimum amplitude, E_{X_L} is at maximum amplitude. Resistive current I_R and voltage E_R are in phase. To find the total current in a resistive-inductive (RL) circuit, the Pythagorean theorem (vector analysis) must be used. Electrically stated, the theorem is $I_T^2 = I_{X_L}^2 + I_R^2$ or $I_T = \sqrt{I_{X_L}^2 + I_R^2}$. In Fig. 45, a current triangle is illustrated to depict current in an RL circuit. Note that I_{X_L} is plotted in the downward direction because it lags E_{X_L}. Calculations, using the values in the figure, are as follows:

(1) *IF I_T IS UNKNOWN* *THEN:* $I_T^2 = I_{X_L}^2 + I_R^2$

GIVEN:

$I_{X_L} = 3$ amps

$I_R = 4$ amps

$$I_T^2 = 3^2 + 4^2$$
$$I_T = \sqrt{9 + 16}$$
$$I_T = \sqrt{25}$$
$$I_T = 5 \text{ amps}$$

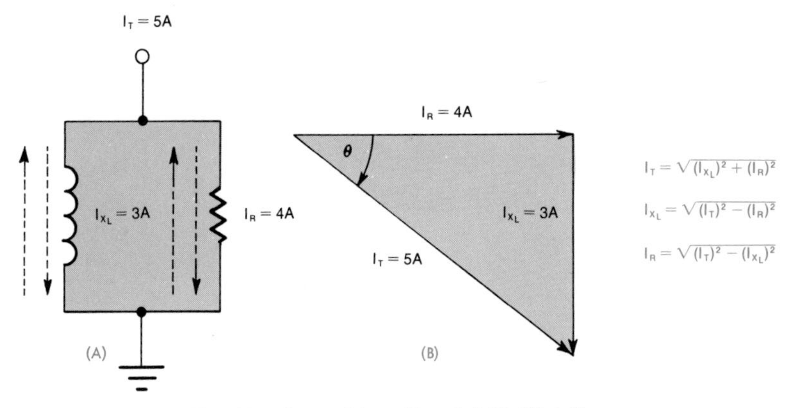

Fig. 45. Current in a Parallel RL Circuit

(2) *IF I_{X_L} IS UNKNOWN* *THEN:* $I_{X_L} = I_T{}^2 - I_R{}^2$

GIVEN:

$I_T = 5$ amps

$I_R = 4$ amps

$$I_{X_L} = 5^2 - 4^2$$
$$I_{X_L} = \sqrt{25 - 16}$$
$$I_{X_L} = \sqrt{9}$$
$$I_{X_L} = 3 \text{ amps}$$

(3) *IF I_R IS UNKNOWN* *THEN:* $I_R = I_T{}^2 - I_{X_L}{}^2$

GIVEN:

$I_T = 5$ amps

$I_{X_L} = 3$ amps

$$I_R = 5^2 - 3^2$$
$$I_R = \sqrt{25 - 9}$$
$$I_R = \sqrt{16}$$
$$I_R = 4 \text{ amps}$$

IMPEDANCE IN A PARALLEL RESISTIVE-INDUCTIVE (RL) CIRCUIT (SEE FIG. 46)

Impedance (Z) in a parallel RL circuit equals the applied voltage E_T divided by the total current I_T. The following calculations are made, using the values shown in Fig. 46.

$$Z = \frac{E_T}{I_T}$$

$$Z = \frac{5 \text{ volts}}{5 \text{ amps}}$$

$$Z = 1 \text{ ohm}$$

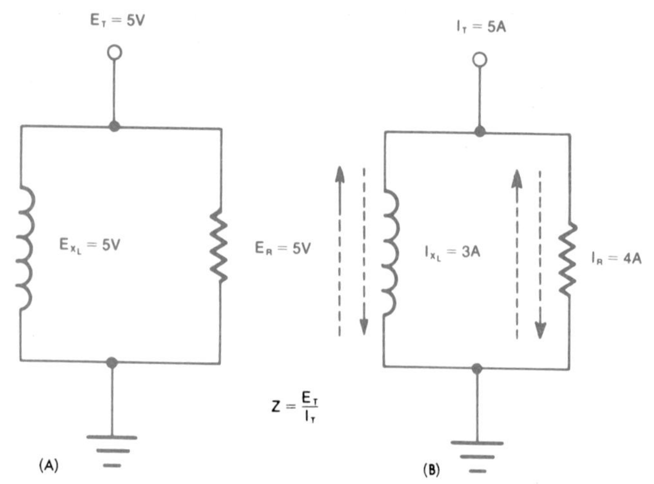

Fig. 46. Impedance in a Parallel RL Circuit

PARALLEL RL CIRCUIT PHASE ANGLES (SEE FIG. 47)

The phase angle in a parallel RL circuit is the relationship between the applied voltage E_T and reactive current I_{X_L}. Since the applied voltage E_T and resistive current I_R are in phase, the reactive current I_{X_L} and resistive current I_R are used for phase angle calculations.

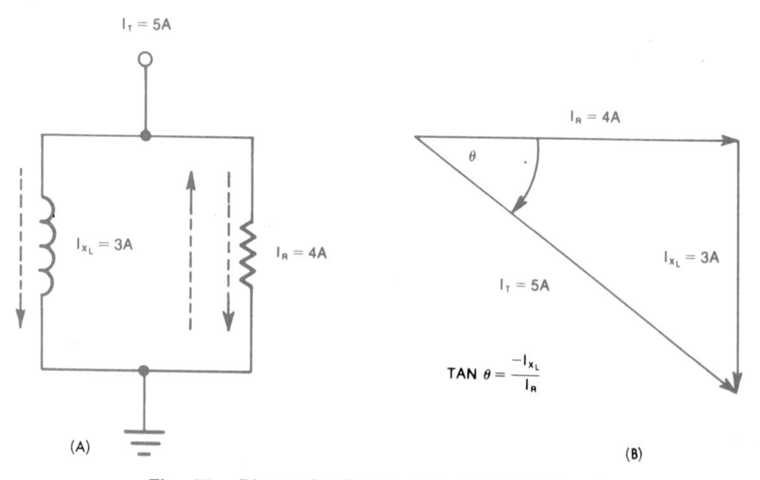

Fig. 47. Phase Angles in a Parallel RL Circuit

In Fig. 47, the reactive current I_{X_L} is plotted downward because I_{X_L} lags E_T by 90 degrees. The sign it takes is negative ($-I_{X_L}$). The following calculations are made using the values in the figure.

$$\text{TAN } \theta = \frac{-I_{X_L}}{I_R}$$

$$\text{TAN } \theta = \frac{-3}{4}$$

$$\text{TAN } \theta = -0.750$$

$$\theta = -36.9°$$

Note: *The phase angle can be calculated in a fraction of a minute with a slide rule of the decimal trig type. Having found this angle θ, it is then extremely quick and easy to find the total current of a parallel RL circuit because it is represented as the hypotenuse of a right triangle. These methods are explained in detail in Appendix D: Solving Electrical Problems with Trigonometry Tables or the Slide Rule. Users of slide rules may prefer the "trig" solution to using the root-mean-square equation because it avoids the use of squares and square roots.*

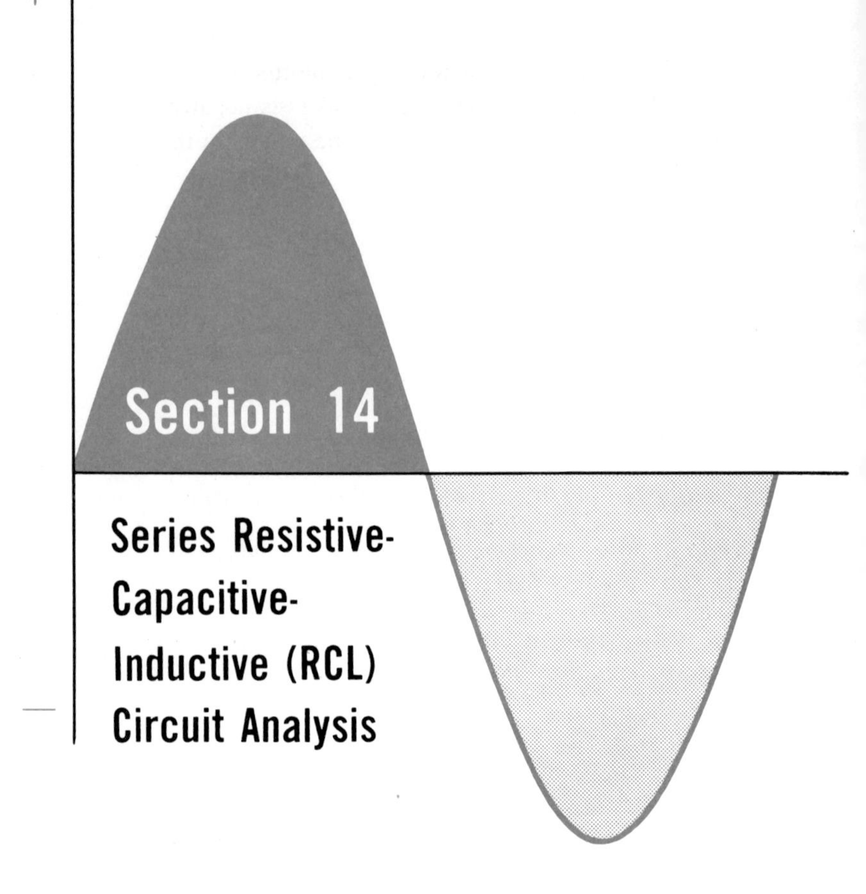

Section 14

Series Resistive-Capacitive-Inductive (RCL) Circuit Analysis

Reaction by both capacitors and inductors are seen in most complex electronic systems. Because the job of signal development is better done with two reacting devices, the two are used to complement one another. Uses of the combination RCL series circuits are innumerable. Nearly every electronic system requires these combinations. Therefore it would be a misnomer to pinpoint one application as *most* useful.

VOLTAGE IN SERIES RESISTIVE-CAPACITIVE-INDUCTIVE (RCL) CIRCUITS (SEE FIG. 48)

Although the current flow through each component of a series RCL circuit is the same, the voltage drop varies with the resistance and the reactance. The total voltage is the vector sum of the resistance voltage E_R and the reactance voltage.

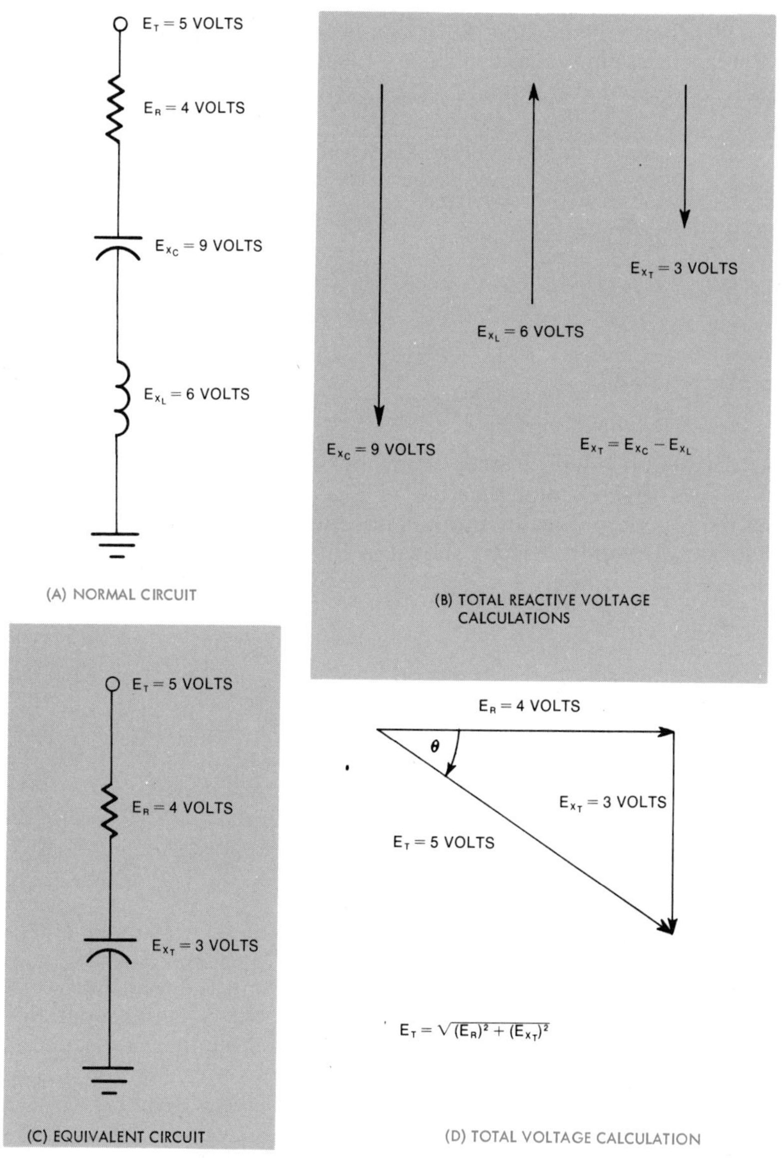

Fig. 48. Voltage in a Series RCL Circuit

Since the capacitive voltage drop and the inductive voltage drop oppose each other, the total reactive voltage drop is the difference between the two or $(E_{X_C} - E_{X_L})$ with the larger of the two becoming the total reactive voltage E_{X_T}.

The total voltage in an RCL circuit is calculated using the Pythagorean theorem (vector analysis). Electrically stated the formula is

$$E_T^2 = E_R^2 + (E_{X_C} - E_{X_L})^2, \text{ or } E_T = \sqrt{E_R^2 + (E_{X_C} - E_{X_L})^2}$$

Calculations, using the values shown in Fig. 48, are as follows:

$$E_T^2 = E_R^2 + (E_{X_C} - E_{X_L})^2$$
$$E_T^2 = 4^2 + (9 - 6)^2$$
$$E_T^2 = 4^2 + 3^2$$
$$E_T = \sqrt{16 + 9}$$
$$E_T = \sqrt{25}$$
$$E_T = 5 \text{ volts}$$

Calculation may also be made by subtracting the two reactive voltages to find total reactive voltage E_{X_T}, then by using Pythagorean theorem (vector sum) as in series calculation for RC or RL circuits. Using the values shown in Fig. 48, the calculations are as follows:

$$E_{X_T} = E_{X_C} - E_{X_L}$$
$$E_{X_T} = 9 - 6$$
$$E_{X_T} = 3 \text{ volts}$$

THEN: $E_T^2 = E_R^2 + E_{X_T}^2$, or $E_T = \sqrt{E_R^2 + E_{X_T}^2}$

$$E_T = \sqrt{4^2 + 3^2}$$
$$E_T = \sqrt{16 + 9}$$
$$E_T = \sqrt{25}$$
$$E_T = 5 \text{ volts}$$

This second method is preferable because it reduces the size of numbers to be squared and requires only one squaring operation and the extraction of square root of a smaller number. Particularly, it is the preferred method where calculations are made on a slide rule, which cannot be used for addition or subtraction.

CURRENT IN A SERIES RESISTIVE-CAPACITIVE-INDUCTIVE (RCL) CIRCUIT (SEE FIG. 49)

Current in series RCL circuits is the same throughout the circuit. The same current flows through the resistor, the capacitor and the inductor. The relationship of resistor current with the resistor volt-

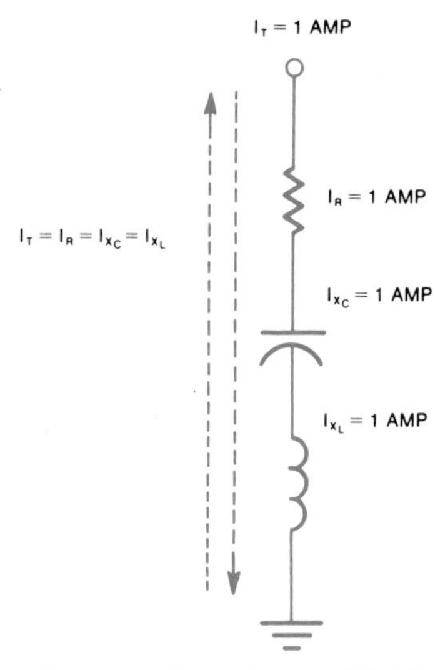

Fig. 49. Current in a Series RCL Circuit

age is in phase. The capacitive and inductive current is out of phase with the capacitive and inductive voltage drops. Total current in a series RCL circuit can therefore be stated:

$$I_T = I_R = I_{X_C} = I_{X_L}$$

IMPEDANCE IN A SERIES RESISTIVE-CAPACITIVE-INDUCTIVE (RCL) CIRCUIT (SEE FIG. 50)

Impedance in a series RCL circuit is the vector sum of the resistance and the reactance. Since the capacitive reactance (X_C) and the inductive reactance (X_L) oppose each other, the difference between the two equals the total reactance (X_T) of the circuit or $X_T = X_C - X_L$.

Impedance is calculated using the Pythagorean theorem (vector

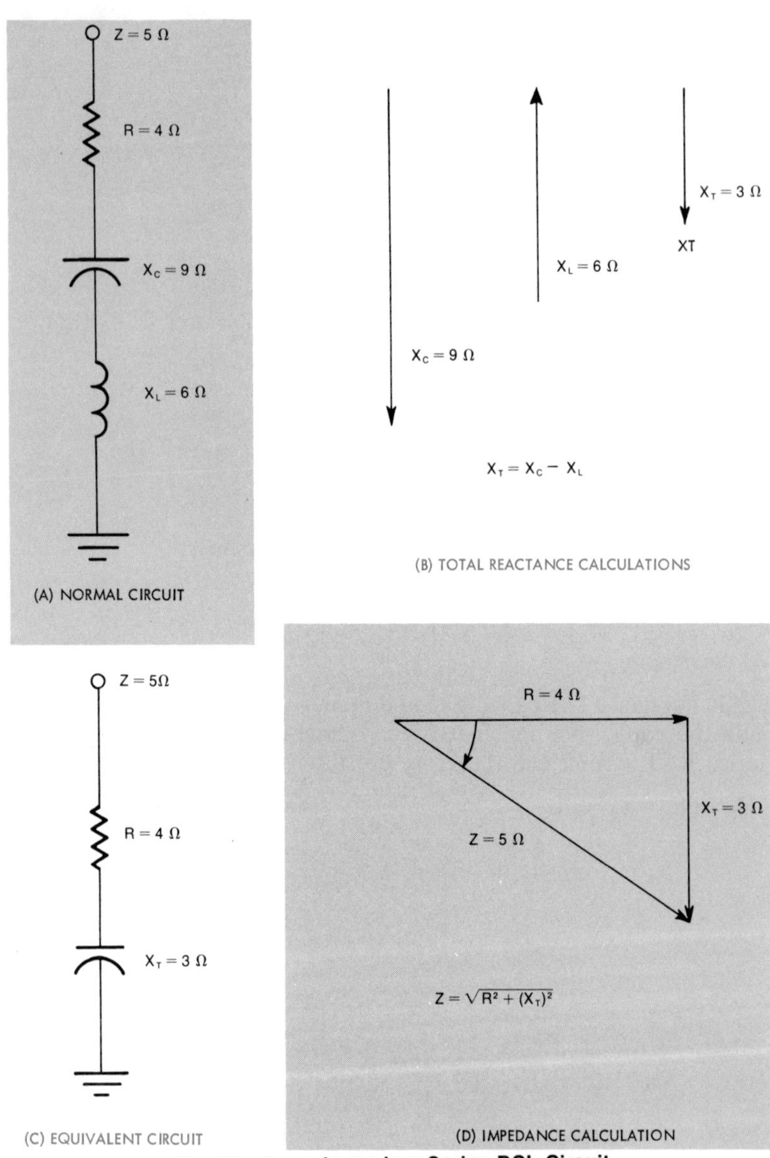

Fig. 50. Impedance in a Series RCL Circuit

analysis). Electrically stated, the formula is $Z^2 = R^2 + (X_C - X_L)^2$ or $Z = \sqrt{R^2 + (X_C - X_L)^2}$. Calculations, using the values shown in Fig. 50, are as follows:

$Z^2 = R^2 + (X_C - X_L)^2$

$Z^2 = 4^2 + (9 - 6)^2$

$Z = \sqrt{16 + 9}$

$Z = \sqrt{25}$

$Z = 5$ ohms

Calculation may also be made by subtracting the reactances to find the total reactance X_T, then by using Pythagorean theorem (vector sum). Using the values shown in Fig. 50, we calculate as follows:

$X_T = X_C - X_L$

$X_T = 9 - 6$

$X_T = 3$ ohms

THEN: $Z^2 = R^2 + X_T{}^2$

$Z^2 = 4^2 + 3^2$

$Z = \sqrt{16 + 9}$

$Z = \sqrt{25}$

$Z = 5$ ohms

This second method is preferable because it reduces the size of numbers to be squared and requires only one squaring operation and the extraction of square root of a smaller number. Particularly, it is the preferred method where calculations are made on a slide rule, which cannot be used for addition or subtractions.

SERIES RCL CIRCUIT PHASE ANGLES

Any RCL circuit can be resolved to an equivalent RC or RL circuit depending on whether capacitive or inductive reactance has the larger value. This is accomplished by subtracting the reactances as shown previously: $X_T = X_C - X_L$.

Phase angle calculations are then the same as those explained under the series RC or series RL circuit phase angle headings.

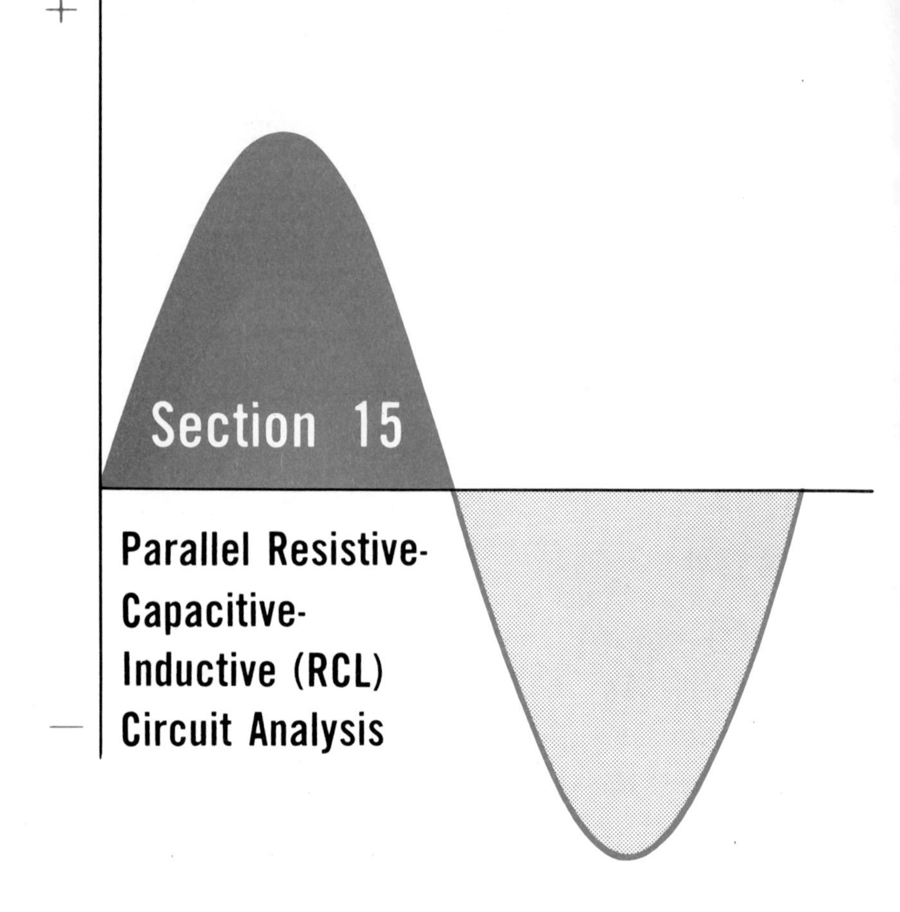

Section 15

Parallel Resistive-Capacitive-Inductive (RCL) Circuit Analysis

Reaction by both capacitors and inductors are seen in most complex electronic systems. Because the job of signal development is better done with two reacting devices, the two are used to complement one another. Uses of the combination RCL parallel circuits are innumerable. Nearly every electronic system requires these combinations. Therefore it would be a misnomer to pinpoint one application as *most* useful.

VOLTAGE IN A PARALLEL RESISTIVE-CAPACITIVE-INDUCTIVE (RCL) CIRCUIT (SEE FIG. 51)

Voltage in parallel RCL circuits is the same on each parallel leg. The same voltage is dropped across the resistor, the capacitor, and the inductor. The relationship of voltage and current through the resistance is in phase. The capacitive and inductive currents are out of phase with the capacitive and inductive voltage drops.

Total voltage in an RCL parallel circuit can therefore be stated: $E_T = E_R = E_{X_C} = E_{X_L}$.

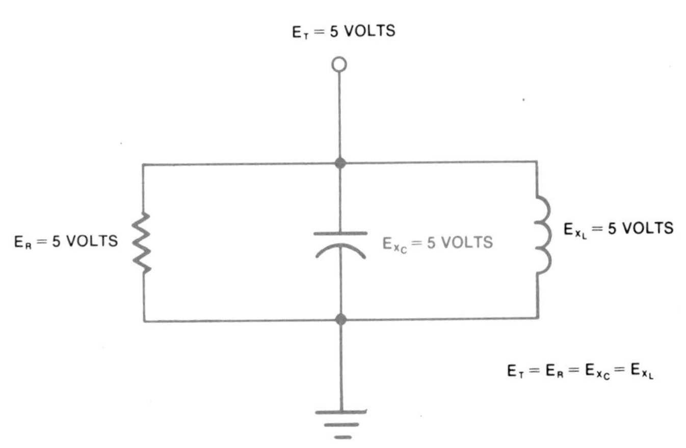

Fig. 51. Voltage in a Parallel RCL Circuit

CURRENT IN A PARALLEL RESISTIVE-CAPACITIVE-INDUCTIVE (RCL) CIRCUIT (SEE FIG. 52)

Although the voltage drop across each parallel leg of a parallel RCL circuit is the same, the current varies within the resistance and reactances.

The total current is the vector sum of the current through the resistance and the current through the reactances. Since the capacitive current and inductive currents oppose each other, the total reactive current is the difference between the two or $(I_{X_C} - I_{X_L})$, with the larger current of the two becoming the total reactive current I_{X_T}.

Total current in a parallel RCL circuit is calculated using the Pythagorean theorem (vector analysis). Electrically stated, the formula is

$$I_T{}^2 = I_R{}^2 + (I_{X_C}{}^2 - I_{X_L}{}^2) \text{ or } I_T = \sqrt{I_R{}^2 + (I_{X_C}{}^2 - I_{X_L}{}^2)}$$

Calculations, using the values shown in Fig. 52 are as follows:

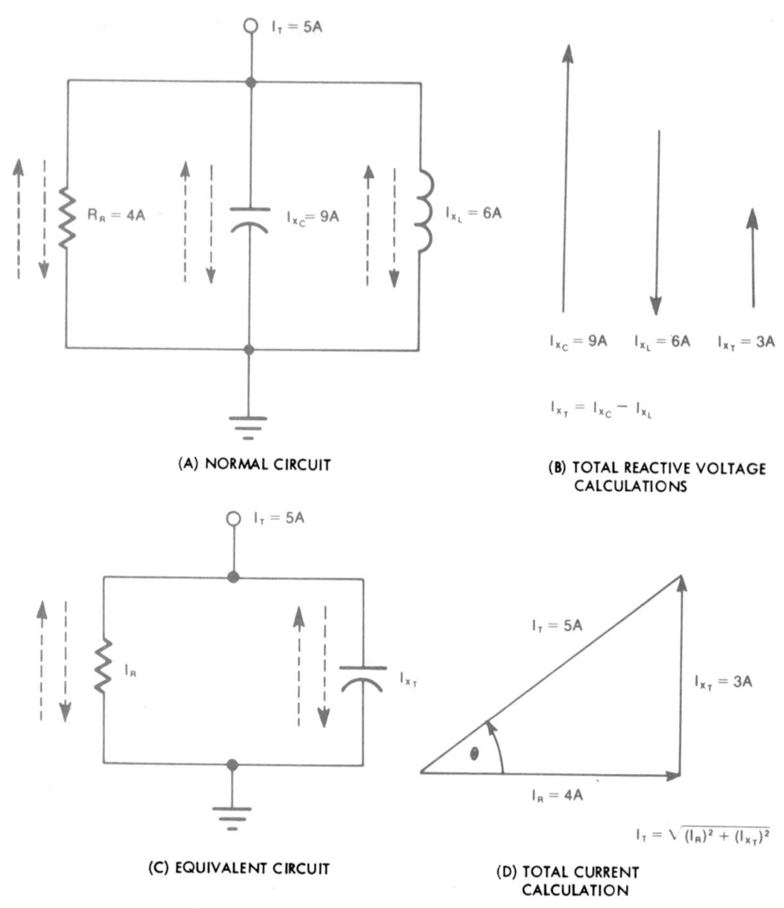

Fig. 52. Current in a Parallel RCL Circuit

$$I_T{}^2 = I_R{}^2 + (I_{X_C} - I_{X_L})^2$$

$$I_T{}^2 = 4^2 + (9 - 6)^2$$

$$I_T{}^2 = 4^2 + 3^2$$

$$I_T = \sqrt{16 + 9}$$

$$I_T = \sqrt{25}$$

$$I_T = 5 \text{ amperes}$$

Calculation may also be made by subtracting the two reactive currents to find total reactive currents I_{X_T}, then by using the Pythagorean theorem (vector sum) as shown in parallel calculations for

RC or RL circuits. Using the values shown in Fig. 52, the calculations are as follows:

$I_{X_T} = I_{X_C} - I_{X_L}$

$I_{X_T} = 9 - 6$

$I_{X_T} = 3$ amps

THEN: $I_T^2 = I_R^2 + I_{X_T}^2$

$I_T = \sqrt{I_R^2 + I_{X_T}^2}$

$I_T = \sqrt{4^2 + 3^2}$

$I_T = \sqrt{16 + 9}$

$I_T = \sqrt{25}$

$I_T = 5$ amps

This second method is preferable because it reduces the size of numbers to be squared and requires only one squaring operation and the extraction of square root of a smaller number. Particularly, it is the preferred method where calculations are made on a slide rule, which cannot be used for addition or subtraction.

IMPEDANCE IN A PARALLEL RESISTIVE-CAPACITIVE-INDUCTIVE (RCL) CIRCUIT

Impedance (Z) in a parallel RCL circuit equals the applied voltage E_T (which is the same through each leg) divided by the total current I_T (which must be added vectorally). The following calculations are made using the values in Figs. 51 and 52, which were previously shown:

$$Z = \frac{E_T}{I_T}$$

$$Z = \frac{5}{5}$$

$Z = 1$ ohm

PARALLEL RCL CIRCUIT PHASE ANGLES

It should be first determined whether the RCL circuit considered is in actuality an RC or an RL circuit. This is accomplished by subtracting the reactive currents as shown previously: $I_{X_T} = I_{X_C} - I_{X_L}$. Phase angle calculations are then the same as those explained under the RC or RL parallel circuit phase angle headings.

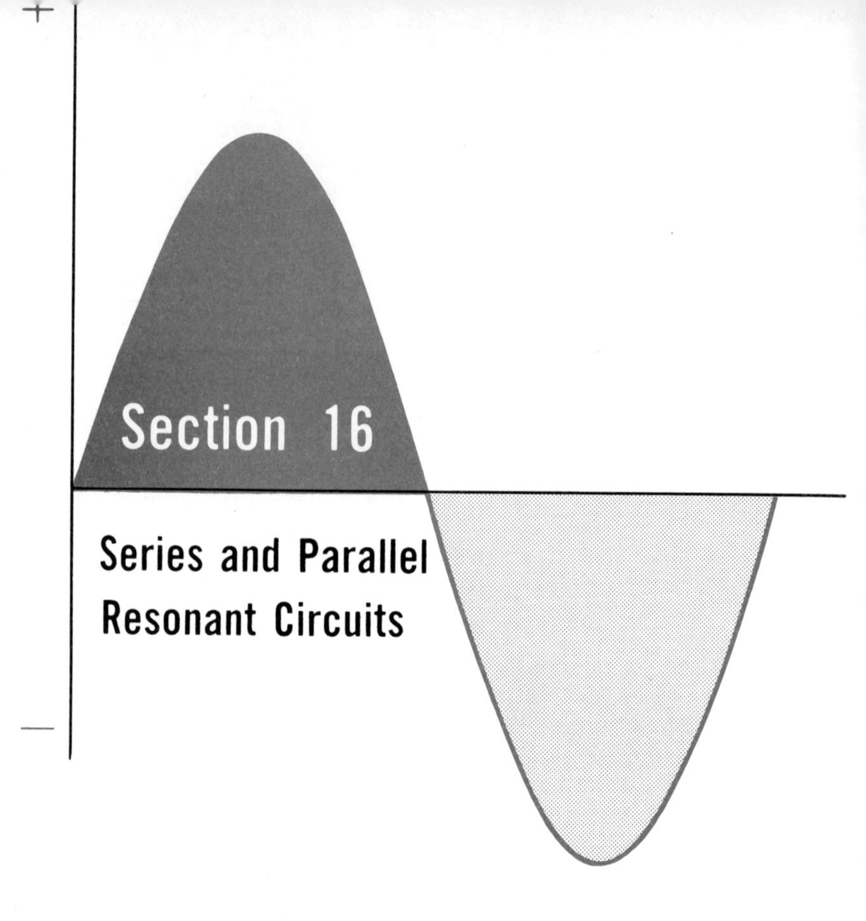

Although resonance is a subject normally confined to tuning or filter circuits, it may be advantageous to discuss it in minor detail at this point. The reason for this is that two major components of ac circuitry (inductors and capacitors) are involved in resonance, and at least a short discussion of their involvement seems necessary.

SERIES RESONANCE

When an inductor and a capacitor are tied in series with a variable frequency source of alternating current, the circuit is called a series tuned circuit.

At low frequencies the reactance of the circuit is primarily capacitive, while at high frequencies the reactance of the circuit is primarily inductive.

At some point in between the high and low frequencies, when the capacitive reactance is equal to the inductive reactance, the circuit is said to be at *resonance* (at resonant frequency). The series circuit is tuned to this frequency. Calculations for finding resonant frequency are as follows:

EXAMPLE:

Let f_r = Resonant frequency

$L = 160$ microhenrys (Inductance)

$C = 160$ picofarads (Capacitance)

$\pi = 3.1416$ (approx.)

$2\pi = 6.28$ (approx.)

THEN: $f_r = \dfrac{1}{2\pi\sqrt{LC}}$

$$f_r = \frac{1}{6.28\sqrt{160 \times 10^{-6} \times 160 \times 10^{-12}}}$$

$$f_r = \frac{1}{6.28\sqrt{(160)^2 \times 10^{-18}}}$$

$$f_r = \frac{1}{6.28 \times 160 \times 10^{-9}}$$

$$f_r = \frac{1}{1004.8 \times 10^{-9}}$$

$f_r = 995{,}223$ hertz (approx. 1000K hertz or 1 megahertz)

Resonant reactance can be calculated by the use of two formulas, the inductive reactance formula and the capacitive reactance formula. Inductive reactance calculations are as follows:

EXAMPLE:

$$X_L = 2\pi fL$$

$$X_L = 6.28 \times 995{,}223 \times 160 \times 10^{-6}$$

$$X_L \cong 1000 \text{ ohms}$$

Capacitive reactance calculations are as follows:

EXAMPLE:

$$X_C = \frac{1}{2\pi fC}$$

$$X_C = \frac{1}{6.28 \times 995{,}223 \times 160 \times 10^{-12}}$$

$$X_C = \frac{1}{0.00104800007040}, \text{ or approx. } \frac{1}{0.001}$$

$$X_C \cong 1000 \text{ ohms}$$

PARALLEL RESONANCE

Parallel resonant frequency calculations are the same as series resonant calculations.

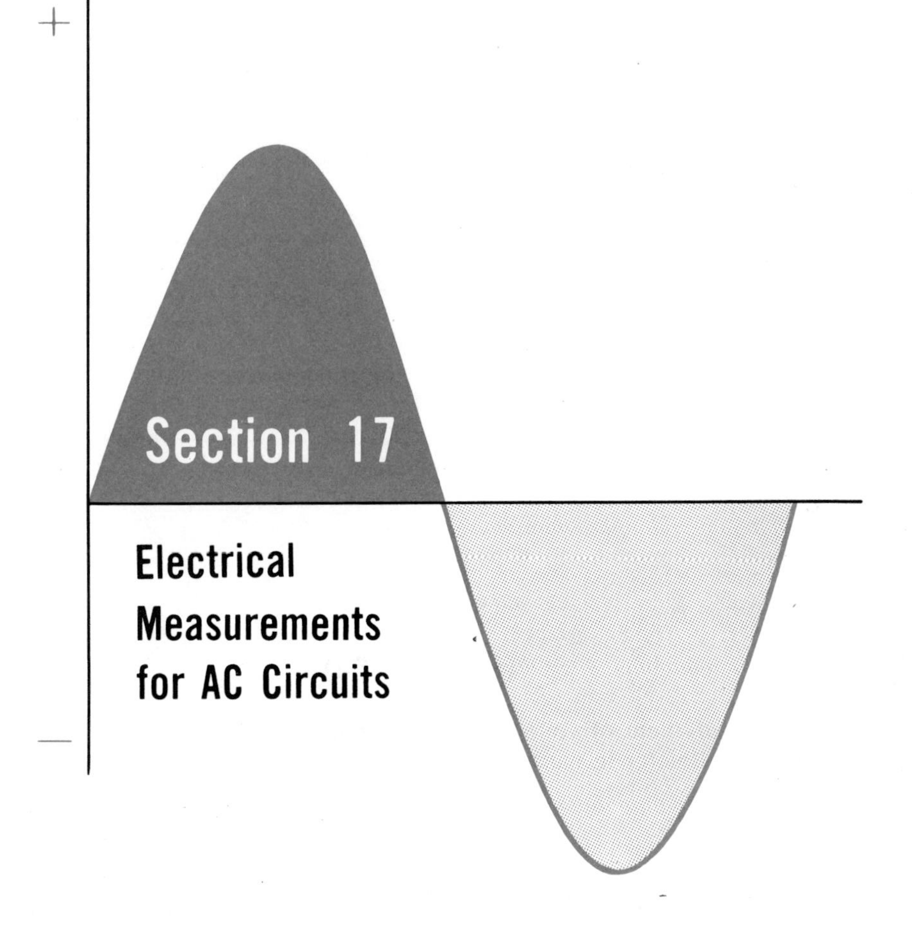

Section 17

Electrical Measurements for AC Circuits

The one major item in the field of electronics that seems to get the least coverage, yet is all-important, is measurement. The accuracy of the measurement depends entirely on the accuracy of the meter you are measuring with. Furthermore, the more accurate the meter, the more expensive it is. Since this is the case, the procedures for measuring should be closely adhered to.

HOW TO MEASURE CURRENT IN AN AC CIRCUIT (SEE FIG. 53)

Current measurements in any ac circuit must be made with an ac ammeter placed in series with the leg to be measured.

In Fig. 53, the ammeter is placed between points *A* and *B* in the series circuit. The ammeter could also be placed between points *C* and *D* or *E* and *F* without changing the ammeter indication.

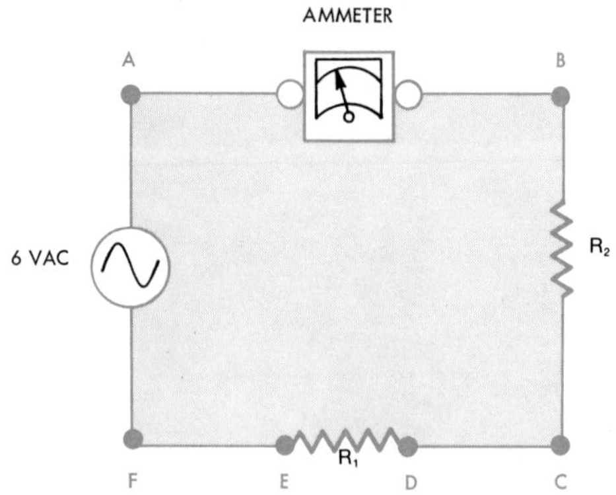

Fig. 53. Measuring Current in an AC Circuit

HOW TO MEASURE VOLTAGE IN AN AC CIRCUIT (SEE FIG. 54)

Measurement of voltage in an ac circuit is performed by placing an ac voltmeter in parallel across the resistance that is to be measured. The voltmeter measures the difference of potential between two points.

In Fig. 54, you will note that the voltmeter is placed across the resistance on points *B* and *C*. From these points the voltmeter measures the voltage drop across the resistor R_2. If the voltmeter

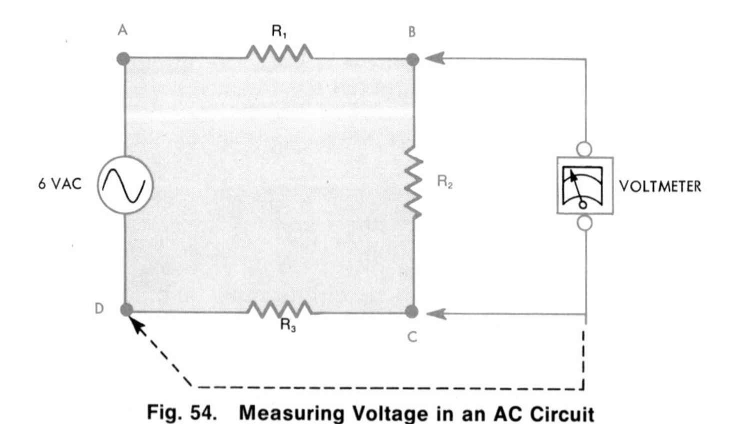

Fig. 54. Measuring Voltage in an AC Circuit

leads were placed on points *B* and *D* as shown by the broken line, the voltage drop across the resistors R_2 and R_3 would be measured. That is, the difference of potential (voltage) between the two points is measured.

HOW TO MEASURE RESISTANCE IN AN AC CIRCUIT (SEE FIG. 55)

Resistance is measured with the ohmmeter leads connected across the resistance to be measured. This places the ohmmeter in parallel with the resistance as shown in Fig. 55.

Resistance of a resistor internal to a circuit cannot be measured efficiently because the rest of the circuit influences resistance, preventing accurate measurement. If it is necessary to measure a resistance in a circuit, remove power from the circuit. Then cut or unsolder one of the resistor leads from the circuit. Then accurate resistance measurement can be made.

Unlike the ammeter and voltmeter, the ohmmeter has its own power supply.

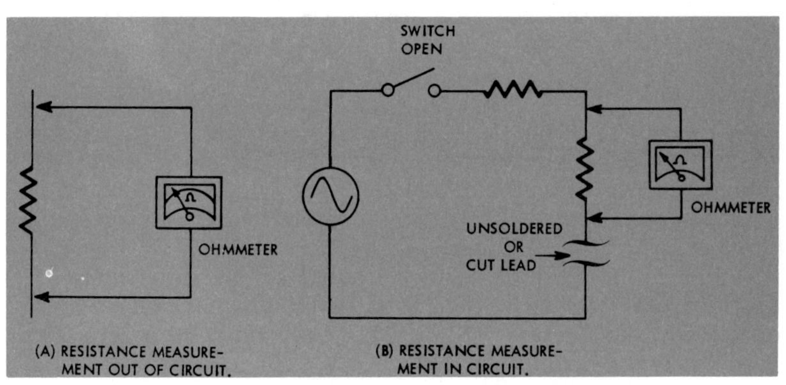

Fig. 55. Measuring Resistance in an AC Circuit

HOW TO MEASURE CONTINUITY IN AN AC CIRCUIT (SEE FIG. 56)

Continuity is a measure of wire resistance. Since a wire has very little resistance, placing ohmmeter leads on its extremities produces

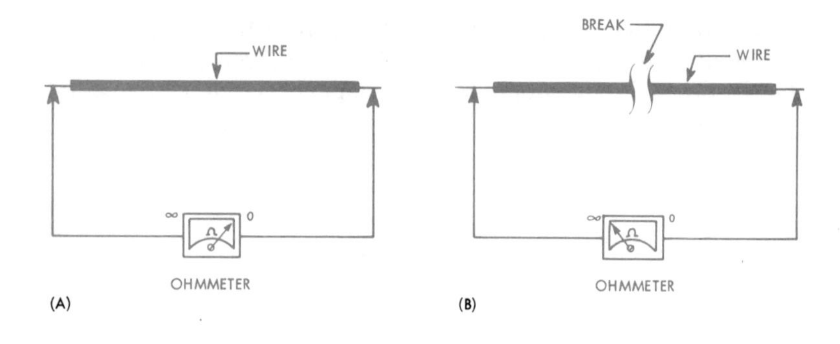

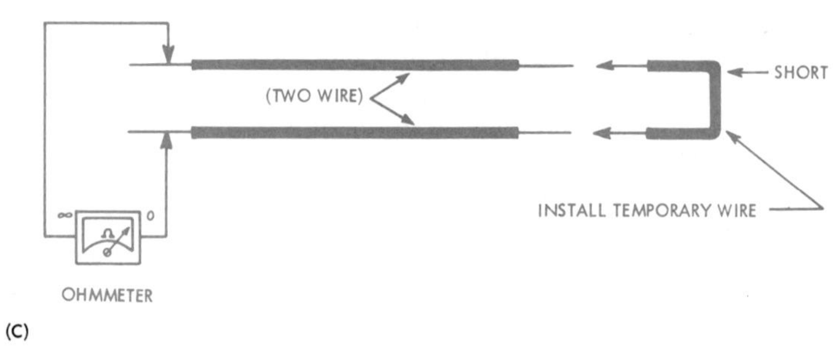

Fig. 56. Measuring Continuity in an AC Circuit

zero (0) or very low resistance, as shown in Fig. 56(A). A low scale should be selected on the ohmmeter.

In Fig. 56(B) the open in the wire causes a large resistance measurement.

In Fig. 56(C), continuity is measured by placing a temporary short on two wire ends, then placing the ohmmeter leads on the other ends. This action will also show continuity by a zero (0) or very low resistance.

APPENDIX A

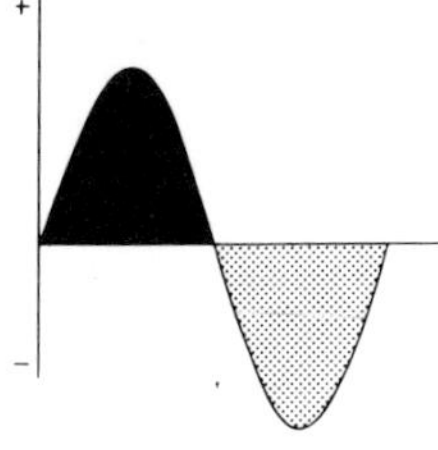

Electrical/Electronic Safety

GENERAL

Safety education today has become an important phase of every training program. Under the 1970 Federal *Occupational Safety and Health Act* (OSHA), the employer is required to furnish a place of employment free of known hazards likely to cause death or injury. The *employer* has the specific duty of complying with safety and health standards as set forth under the 1970 act. At the same time, *employees* also have the duty to comply with these standards.

A full treatment of the subject of safety is far beyond the scope of this book, and there is ample justification for a full course in safety procedures, including first aid treatment, for all electrical and electronic technicians. Instructors should be certified by the state and qualified for any special electrical applications. The intent of this chapter is to make students fully aware of the ever-present, invisible, and generally silent hazards in handling electrical apparatus and to point out some fairly common causes of electrical shocks and fires that can easily be overlooked.

Since the time when Benjamin Franklin flew his famous kite it has become more and more apparent that electricity, even in its milder forms, is *dangerous*. A fact not widely mentioned in history is that shortly after Franklin's experiment a Russian experimenter was killed in his attempt to duplicate the kite trick. We can therefore assume that good safety habits are mandatory for all who use, direct the use of, or come in contact with electricity. Electrical equipment is found in every place that the ordinary person may find himself. For this reason it becomes the responsibility of each of us to be knowledgeable of electrical safety and to become his own and his brother's keeper.

The most basic cause of electrical accidents, as other types of accidents, is *carelessness* and the best prevention is *common sense*. However, knowing exactly what to do in an emergency is only achieved through formal education or experience. Unfortunately, the experience could be fatal, so it is more desirable to derive your knowledge through schooling. When an emergency happens it is often accompanied by panic that can cause the mind or muscles to become paralyzed. The antidote for this panic is education.

RESPONSIBILITY

Electrical/electronic safety is especially important to the technician who is exposed to electrical equipment in the raw state. He is likely to be the one who comes into direct contact with electrical and/or electronic components that may do harm to the body. Since this is the case, he finds himself in the precarious position of being continually alert to hazards that may affect him, his associates, and the people who use the equipment he builds or repairs. The supervisor is responsible for enforcing the rules of safety in the area under his direct supervision. Inspectors ensure that equipment is tested before it is released for use, and final testing should include safety precautions. Finally, the user of the equipment should be qualified in its operation so he may know when it is operating in the proper manner. *No one* who comes into contact with such equipment is exempt from some degree of responsibility.

The average layman may not believe he can get a severe shock from an electrical component disconnected from a power source. The electrical or electronic technician should know better. You know from your study of capacitance, for instance, that capacitors hold electrical charges which they later discharge. A fully charged capacitor disconnected from the power source can deliver a severe and possibly fatal shock, so you take the precaution of "shorting" it before removing it from the equipment. Also, as your brother's keeper, you should warn the user of this hazard.

ELECTRICAL SHOCK

There is a common belief that it takes a great volume of electricity to cause a fatal shock and that high voltage is the thing that provides the jolt to do the job. Imaginative stories about people being literally fried or jolted from their shoes are supplied freely by storytellers. Although there are true tales of this type, this is not generally the case. The bulk of electrical shocks come in small packages. Death from electrical shock happens most often from ordinary 60-hertz, 115 VAC house power. The effect, in general, is an instantaneous, violent type of paralysis. The human body contains a great deal of water and is normally somewhat acid and saline. For these reasons it is a fairly good conductor of electric current, which has no regard for human life or feelings.

Our brains are loaded with tiny nerve ends that provide transmitting service to the muscles. Muscles, in turn, provide motion for the body functions such as the heart and lungs. Now, assume that an electrical impulse that was not called for told your heart to speed up its pumping job or to stop its pumping job. Or suppose that a similar impulse of electricity told the lungs to quit taking in air. These situations do occur and with a comparitively small amount of current flow. Depending on a multitude of complex variables, a small current such as 10 milliamps can be very unpleasant. Currents no larger than 20 milliamps can cause muscle tightening or freezing. Currents of 30 milliamps can cause damage to brain tissue and blood vessels. And damage to brain and blood vessels can, of course, be fatal.

As you know, a decrease in resistance (according to Ohm's Law) causes current to increase. Body resistance decreases when perspiring or otherwise wet. A great number of things can cause variation in resistance. The general health of the person is probably the most important variable. A body in good condition has a much better chance of recovering from electrical shock. The muscles, by being in better tone, can recover to normal from paralyzation.

The path of current flow varies the shock. For instance, if the current involves the brain or heart it is naturally more dangerous than another path. The length of exposure can also be a factor, as well as the size or surface area of the electrical contact. Large voltages can cause spastic action, but recovery can be rapid. Also, small currents can cause muscle paralysis.

In the event of paralysis, artificial respiration or massaging must take place as quickly as possible to prevent loss of body functions and damage to the brain because of lack of oxygen. A condition known as ventricular fibrilation (uncoordinated heart beats, both fast and irregular) can occur with high currents, say 50 to 100 milliamps. This action will continue until something is done externally to restore regular heartbeat.

Death normally occurs when currents reach 250 milliamps. This does not have to occur, however, as rapid first aid can save a victim whose heart has actually stopped beating or whose lungs have stopped pumping momentarily. In many cases rapid action by knowledgeable persons can prevent body damage and save lives.

RAPID RESCUE TECHNIQUES FOR ELECTRICAL EXPOSURE

A special course in first aid will ensure the proper methods. A wrong method can be worse than no action at all. The first rule of thumb for an electrical shock victim is to remove the current path. This can be done by turning the power switch off if it is readily accessible. If not accessible, the person can be detached from the current source by using an insulator of some sort such as wood, rubber, cork, or plastic. Sometimes it is more suitable and sensible to remove the power source from the victim. Whichever is the case, isolation from the current source is by far the most important and first move that can be made.

Isolation procedures can cause problems. The rescuer may find himself in the current path. Touching a person who is paralyzed to a current source provides the current source another path through the rescuer's body. Care must be taken to prevent this from happening. After isolation, artificial respiration and/or other first aid should be applied. In all cases speed is vital. Death occurs in direct proportion to time. It is therefore obvious that a person given artificial respiration in the first three minutes has a much better chance of survival than one who is given artificial respiration after five minutes.

SNEAKY ELECTRICAL CONDUCTORS

In every electrical activity there are sneaky conductors of electricity that cause continuous problems of "shorts" and therefore electrical shock. Anywhere that you have electricity it is not just the wire, the source, or the load that provides these paths for current flow. For instance, cement may seem dry and clean but have moisture in it. In this condition the cement could be a sneaky conductor. Metal floors are, of course, good conductors. A sweating body can cause a multitude of problems. Machines of all types in the general area will serve as conductors. Steel building posts can be conductors as well as metal roofs, steel desks, pots and pans, bicycles, automobiles, refrigerators, washing machines, and just about any other metal object you may name. These sneaky conductors can cause current draw.

Make sure, then, that you are properly grounded or that the equipment you are working with or near is grounded. Use insulated tools to prevent current from flowing where it isn't supposed to. Use floor pads and keep water and oil from floors around which you are performing electrical work. Water does not mix with electricity. Keep debris and scraps picked up to avoid similar situations. Cleanliness and alertness will help avoid or eliminate sneak circuits.

BATTERY HANDLING

Three basic safety problems are associated with the handling of batteries. These are: acid burns, fires, and explosions. Acid burns may be prevented by use of battery-handling clothes and equipment. Clothes for this purpose are mostly made of rubber and include such things as aprons, gloves, boots, and special glasses.

Proper tools are essential to perform the correct procedures safely. Proper flooring will prevent falling and spilling acids.

Fire and explosion may be caused by ignition of gases given off from the charging action. These gases, when mixed with air (oxygen), provide a highly flammable and explosive situation. Gases should not be allowed to accumulate. Ventilators should be installed in battery shops to expel the dangerous gases. Smoking in the area should be prohibited. Signs should be installed warning everyone who might enter the battery shop of the dangers that are within.

HOW TO CONTROL AN ELECTRICAL FIRE

An electrical fire is caused by current flow to some circuit that cannot withstand the current level. Also, electrical fires are caused by sneak circuits which accidently draw current, for instance a "short" to the case in a motor-driven furnace. In any event, since the cause of the problem is current flow, disconnecting the current should be the first step in eliminating the problem. Remove power from the circuit preferably by throwing a switch or by isolating the fire, using insulating material such as wood or plastic. Cut wires with wooden handled hatchets or some similar device. Prevent yourself from becoming part of the circuit. After removing the current, call the fire department, then put out the fire.

Electrical fires are best extinguished with the use of carbon dioxide (CO_2) directed toward the base of the fire. Do not use foam, as it conducts electricity.

GOOD SOLDERING HABITS

Soldering irons or soldering guns all have one thing in common: they are hot. Each is hot enough to melt solder joints. The actual temperature varies with the solder type. The speed at which the soldering iron or gun melts the solder joints is dependent on the wattage of the iron or gun and the size or complexity of the joint. In all events, the soldering device must not only be protected from the handler but also from the other circuits or equipment around it.

The soldering iron or gun should be placed in a heat-sink holder between soldering actions. Heat sinks should also be used to protect electrical/electronic circuit components. Danger of work contamination is always present as dripping or stringing of solder may occur during soldering operations. Fire hazards are always present when working with heat.

Electrical fire hazards may be prevented by ensuring that power is removed from equipment being worked on. Soldering operations should take place only after proper preparation of the work area. Clean and dry work areas, the proper wattage iron, and a well laid out soldering plan help prevent soldering accidents.

Appendix B

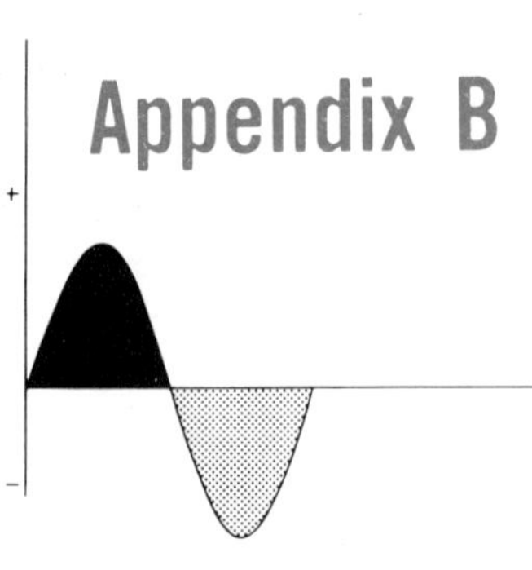

Calculating with Powers of Ten

Electricity, including electronics, is unusual in that electrical units and calculations are now and always have been metric, even in the United States. This is because work in electricity originated in Europe, where it developed out of science rather than gradually out of previous experience. Scientists everywhere have used metric units of one kind or another for over 150 years because they found the older traditional units too cumbersome and poorly related. However, the electrical worker in the United States (unlike about 92 percent of the world, which uses metric units for *all* purposes) must still use traditional U.S. units where electricity interfaces with mechanical engineering, manufacturing, building trades, and other occupations and practices.

While the metric units used in electrical work are of great advantage in avoiding computation with common fractions, there is another problem to be faced and overcome. In electrical and electronic calculations the size of numbers commonly used extends over a tremendous range—from microscopic to astronomical, in fact.

ORDINARY DECIMAL NOTATION INADEQUATE

Ordinary decimal notation, quite adequate for many practical purposes, is much too cumbersome for electrical calculations. The formulas and equations used are generally stated in base units for simplicity. In practice, however, many of these base units (such as the ampere, the volt, the ohm, the farad, the henry, and the watt) are either much too large or much too small for given applications.

For instance, electric power lines may transmit several million volts (megavolts), while an electronic circuit may be sensitive to thousandths of a volt (millivolts). Similarly, while the base unit of current (ampere) is a satisfactory unit for most household appliances, thousandths of an ampere (milliamperes) or even millionths of an ampere (microamperes) are common in electronic circuits. Frequencies in hertz (cycles per second) extend from zero (in a pure dc circuit) to about 300 billion hertz (gigahertz or GHz) for weather radar and NASA communications. The base unit for capacitance is the farad, but more commonly millionths of a farad (microfarads) or even trillionths of a farad (picofarads) are more convenient for electronic work.

Consider the fairly simple equation for capacitive reactance (X_C) with frequency (f) in hertz (Hz), capacitance (C) in farads (F) and capacitive reactance (X_C) in ohms (Ω):

$$X_C = \frac{1}{2\pi\, fC} \quad \text{(the constant } 2\pi = 6.28 \text{ approx.)}$$

Now suppose that X_C is unknown, f = 200 MHz (that is 200,000,000 Hz), and capacitance is 372 μF (that is 0.000372 F). In the ordinary decimal notation this simple equation becomes rather a nightmare:

$$X_C = \frac{1}{6.28 \times 200{,}000{,}000 \times 0.000372}$$

$$\cong 0.00000214 \text{ ohm, or } 2.14\ \mu\Omega \text{ (microhms)}$$

Handling numbers in this way would be extremely tedious, and errors could easily be made. Such calculations are relatively easy, however, by the "powers of ten" method.

THE POWERS OF TEN NOTATION AND METRIC PREFIXES

There is nothing at all difficult about the powers of ten notation, but it does take some practice to use it with confidence. It is simply an extension of the familiar decimal system and uses the laws of exponents applied to the base 10 to obtain the "powers of 10" that you will learn to use as multipliers. This method is indispensable for technical calculations of all kinds in either metric or customary U.S. units, but is especially adaptable to metric units. The reason is that the metric system uses prefixes that name the powers of ten used as multipliers. This is shown in Table B-1.

Positive powers of ten are known in metrics as *multiples* of the base unit and are always larger than the base unit. Thus the prefix "kilo" in kilometer means "thousand", which is 10^3 or 10 to the third power. The kilometer, a thousand times longer than the meter, is a multiple of the meter.

Negative powers of ten, always written with the minus sign before the exponent, are known as *sub-multiples* of the base unit

TABLE B-1. STANDARD PREFIXES, THEIR SYMBOLS AND MAGNITUDES

FACTOR BY WHICH UNIT IS MULTIPLIED	PREFIX	SYMBOL	OLD PREFIX & SYMBOL (IF DIFFERENT)	
10^{12}	tera-	T		
10^{9}	giga-	G	kilomega-	km
10^{6}	mega-	M		m
10^{3}	kilo-	k		
10^{2}	hecto-	h		
10^{1}	deka-	da		
10^{-1}	deci-	d		
10^{-2}	centi-	c		
10^{-3}	milli-	m		
10^{-6}	micro-	μ		
10^{-9}	nano	n	millimicro-	$m\mu$
10^{-12}	pico-	p	micromicro-	$\mu\mu$
10^{-15}	femto-	f		
10^{-18}	atto-	a	KI	

EXAMPLES:

Gigahertz = GHz	Millisecond = ms
Megohm = MΩ	Microhenry = μH
Kilovolt = kV	Picofarad = pF
Centimeter = cm	

and are always smaller than the base unit. Thus the prefix "milli" in millimeter means "thousandth", which is 10^{-3} or 10 to the negative third power. The millimeter, a thousand times shorter than the meter, is a sub-multiple of the meter. Notice, however, that using a negative power does *not* result in a negative number.

It will be worth your while to learn the powers of ten along with their associated metric prefixes and the letter symbols for these prefixes. Learn these well enough to recall them instantly. Then you are ready for the next step, which is how to express any number, large or small, in the powers of ten notation.

EXPRESSING ANY NUMBER IN THE POWERS OF TEN NOTATION

A number of any magnitude in the ordinary notation can be expressed as a single digit between 1 and 10, multiplied by a power of ten. Usually the single-digit whole number will be followed by a decimal part. For instance, 1.734 is actually a number between 1 and 2.

Taking two examples for illustration, the number 1,734,000 may be written as 1.734×10^6, and similarly the number 0.001734 may be written as 1.734×10^{-3}.

Notice that in either case the *coefficient* 1.734 is the same, because both numbers have the same digits in the same sequence. They differ only in the power of ten used: 10^6 for the large number and 10^{-3} for the small number in these examples. A point to notice is that any number which is entirely decimal in the ordinary notation, such as 0.83 or 0.0000083, is less than 1, and in the powers of ten notation will have a negative power of ten as the multiplier: in the first case $0.83 = 8.3 \times 10^{-1}$ and in the second case $0.0000083 = 8.3 \times 10^{-6}$.

In these illustrations, expressing the number as basically a single digit between 1 and 10 times the appropriate power of ten, the resulting number, such as 1.734×10^6, is in a form known as *scientific notation.* This is a very solemn name for a very simple thing. But it is not necessary to express numbers in the scientific notation form. For one reason or another, in your calculations it may be more convenient to have a different number of digits as whole numbers in the coefficient part of the expression. This is easy. 1.734×10^6 may also be written as 17.34×10^5, for instance.

To reduce the number of digits of large numbers, move the decimal point to the left as far as desired, then count the number of positions to the original decimal point. This count gives the proper exponent for the power of ten.

EXAMPLES:

(a) $470{,}000 = 0.47 \times 10^6$, or 4.7×10^5, or 47×10^4
(b) $31{,}416 = 314.16 \times 10^2$, or 3.1416×10^4, or 0.31416×10^5
(c) $285 = 2.85 \times 10^2$, or 0.285×10^3, or 0.0285×10^4

Often it becomes necessary to change numbers expressed in powers of ten back to the ordinary notation. An example is the approximate speed of light or of radio frequency waves, stated scientifically as 1.86×10^5 miles per second. Here the fact that no significant digits are shown beyond the third implies that this speed is known only to that degree of accuracy. In common notation this is 186,000 miles per second. This conversion is made by reversing the process previously shown: the decimal is moved back (toward the right) 5 places, as indicated by the exponent 5 in the power of ten. A little practice is recommended.

EXAMPLES:

(a) $0.47 \times 10^6 = 470{,}000$, or $4.7 \times 10^5 = 470{,}000$
(b) $314.16 \times 10^2 = 31{,}416$, or $3.1416 \times 10^4 = 31{,}416$
(c) $2.85 \times 10^2 = 285$, or $0.285 \times 10^3 = 285$

Up to this point all the numbers were larger than 1 as expressed in common notation or as positive powers of ten in the powers of ten notation. This brings us to numbers less than 1, which are the entirely decimal numbers in common notation. In the powers of ten notation these small numbers are expressed with negative powers of ten. The procedure is equally simple.

To express numbers smaller than 1 in the powers of ten notation, move the decimal point to the right as far as desired, then count the number of positions to the original decimal point. This count gives the proper negative exponent for the power of ten.

EXAMPLES:

(a) $0.000372 = 372 \times 10^{-6}$, or 37.2×10^{-5}, or 3.72×10^{-4}
(b) $0.004567 = 4.567 \times 10^{-3}$, or 45.67×10^{-4}, or 456.7×10^{-5}
(c) $0.971 = 9.71 \times 10^{-1}$, or 97.1×10^{-2}, or 971×10^{-3}

Again, it may be necessary to change numbers expressed in negative powers of ten back to the ordinary notation. This is accomplished by reversing the previous process: the decimal is moved back (toward the left) the number of positions indicated by the negative exponent of the power of ten. Zeros may have to be inserted. A little practice in this is recommended.

EXAMPLES:

(a) $372 \times 10^{-6} = 0.000372$, or $37.2 \times 10^{-5} = 0.000372$
(b) $4.567 \times 10^{-3} = 0.004567$, or $456.7 \times 10^{-5} = 0.004567$
(c) $1.293 \times 10^{-3} = 0.001293$, or $12.93 \times 10^{-4} = 0.001293$

One final note completes the discussion of powers of ten notation. In making calculations with powers of ten, there are occasions to subtract one exponent from another, as for example 10^{6-6}, giving 10^0. This is puzzling to many people because it doesn't appeal to common sense and seems to mean nothing. Mathematically, however, the zero power of any number, large or small, is 1. So a quantity such as $2.52 \times 10^0 = 2.52 \times 1 = 2.52$. The zero power of ten simply disappears.

USING POWERS OF TEN IN ACTUAL CALCULATIONS

The main processes of calculation which concern us are:

1. Addition
2. Subtraction
3. Multiplication
4. Division
5. Reciprocals
6. Combined Operations
7. Squares
8. Square Roots
9. Cubes
10. Cube Roots
11. Quick Approximations
12. The E Notation

Each of these processes will be explained with examples in paragraphs that follow.

1. Addition

Addition of numbers expressed in powers of ten is possible only if the same power of ten is used in all numbers. If this is not the case, first change the powers of ten to a common power. Only the significant digits are then added in the ordinary way, and the addition is multiplied by the common power of ten.

EXAMPLES:

(a)
$$\begin{array}{r} 432 \times 10^6 \\ + \; 18 \times 10^6 \\ \hline 450 \times 10^6 \end{array}$$

(b)
$$\begin{array}{r} 224 \times 10^{-4} \\ + \; 32 \times 10^{-4} \\ \hline 256 \times 10^{-4} \end{array}$$

(c)
$$\begin{array}{r} 1437 \times 10^2 \\ 13 \times 10^2 \\ +3415 \times 10^2 \\ \hline 4865 \times 10^2 \end{array}$$

(d)
$$\begin{array}{r} 286 \times 10^{-6} \\ + \; 14 \times 10^2 \\ \hline \end{array}$$

(Not possible to add)

2. Subtraction

Subtraction of numbers expressed in powers of ten is possible only if the same power of ten is used in both numbers. If this is not the case, first change the powers of ten to a common power. Subtraction of the significant digits is then performed in the ordinary way, and the remainder is multiplied by the common power of ten.

EXAMPLES:

(a) 432×10^6
$- \ 18 \times 10^6$
414×10^6

(b) 224×10^{-4}
$- \ 32 \times 10^{-4}$
192×10^{-4}

(c) 286×10^{-6}
14×10^2

(Not possible to subtract)

3. Multiplication

A mathematical law states that when powers are multiplied their exponents are added. Powers of ten are no exception to this law of exponents. Therefore, in multiplying numbers expressed in powers of ten, the significant digits are multiplied in the ordinary way and their product is then multiplied by the new power of ten obtained by adding the exponents.

With a little practice this becomes easy.

EXAMPLE:

(a) $372{,}000 \times 3{,}200 = 3.72 \times 10^5 \times 3.2 \times 10^3$
$= 3.72 \times 3.2 \times 10^{5+3}$
$= 11.904 \times 10^8$
$\cong 11.9 \times 10^8$ (correct to 3 significant digits)

Note here that in common notation this answer would be written as 1,190,000,000—a huge, unwieldy number.

4. Division

A mathematical law states that when one power is divided by another the exponent of the divisor is subtracted from the exponent of the dividend. Powers of ten are no exception to this law of exponents. Therefore, in dividing numbers expressed in powers of ten, the significant digits are divided in the ordinary way and their quotient is then multiplied by the new power of ten obtained by subtracting the exponent of the divisor power from the exponent of the dividend.

This is actually easier to do than to explain, and with a little practice the process becomes easy.

EXAMPLES:

(a) $300 \div 200 = \frac{3 \times 10^3}{2 \times 10^2} = \frac{3}{2} \times 10^{3-2}$

$= 1.5 \times 10^1$

In common notation the answer is 15.

(b) $\frac{372{,}000}{1{,}860} = \frac{3.72 \times 10^5}{1.86 \times 10^3} = \frac{3.72}{1.86} \times 10^{5-3}$

$= 2 \times 10^2$

In common notation this would be 200.

(c) $112{,}000 \div 0.00028 = \frac{112{,}000}{0.00028}$

$= \frac{1.12 \times 10^5}{2.8 \times 10^{-4}}$

$= \frac{1.12}{2.8} \times 10^{5-(-4)}$

$= 0.4 \times 10^9$

$= 4 \times 10^8$

And this number in common notation is 400,000,000.

(d) $6.45 \div 0.00129 = \frac{6.45 \times 10^0}{1.29 \times 10^{-3}} = \frac{6.45}{1.29} \times 10^{0-(-3)}$

$= 5 \times 10^3$

And this number in common notation is 5,000.

Another rule of importance in division was illustrated in Problem (d) where it was found that $\frac{10^0}{10^{-3}} = 10^3$. *Any power of ten may be moved from the numerator to the denominator or from the denominator to the numerator if its sign is changed.*

EXAMPLES:

(a) $\frac{1}{10} = \frac{1}{10^1} = 10^{-1} = 0.1$

(Actually 10^{-1} is an alternate way of writing 1/10.)

(b) $\frac{1}{100} = \frac{1}{10^2} = 10^{-2} = 0.01$

(c) $0.0002 = 2.0 \times 10^{-4} = \frac{2}{10^4}$, or in common notation 2/10000.

(d) $52{,}000{,}000 = 5.2 \times 10^7 = \dfrac{5.2}{10^{-7}}$

And this number in common notation is $\dfrac{5.2}{0.0000001}$

Now as a practice exercise, start with this last number and reverse these steps.

Another rule in division of powers of ten is that like powers in both numerator and denominator may be cancelled.

EXAMPLES:

(a) $\dfrac{104 \times \cancel{10^5}}{13 \times \cancel{10^5}} = \dfrac{104}{13} = 8$

(The proof is that $10^{5-5} = 10^0$, and $10^0 = 1$)

(b) $\dfrac{344 \times \cancel{10^{-4}}}{86 \times \cancel{10^{-4}}} = \dfrac{344}{86} = 4$

(The proof is that $10^{-4-(-4)} = 10^{-4+4} = 10^0 = 1$.)

5. Reciprocals

The reciprocal of a number is defined as 1 divided by that number. Also, the reciprocal of a number written in fractional form is the number obtained by inverting the fraction. These definitions apply to numbers expressed in the powers of ten notation as well as the common notation.

Difficult reciprocal calculations are common in electrical equations, and these may be solved easily using powers of ten as shown.

EXAMPLE:

$\dfrac{1}{20{,}000 \times 0.000005 \times 0.000025}$ in common notation

$= \dfrac{1}{2 \times 10^4 \times 5 \times 10^{-6} \times 2.5 \times 10^{-5}}$

$= \dfrac{1}{2 \times 5 \times 2.5 \times 10^{4-6-5}}$

$= \dfrac{1}{25 \times 10^{-7}}$

$= \dfrac{10^7}{25} = \dfrac{10^2}{25} \times 10^5 = 4 \times 10^5$

In common notation $4 \times 10^5 = 400{,}000$.

6. Combined Operations

Many electrical equations contain combinations of large and small numbers that are to be divided and multiplied. *By first converting to the powers of ten notation and then performing the indicated calculations by rules you have already learned, you will find these can be readily solved.*

EXAMPLE:

$$\frac{0.000012 \times 8{,}000 \times 12{,}000}{3{,}000{,}000 \times 0.000004} \text{ in common notation}$$

$$= \frac{1.2 \times 10^{-5} \times 8 \times 10^{3} \times 1.2 \times 10^{4}}{3 \times 10^{6} \times 4 \times 10^{-6}}$$

$$= \frac{1.2 \times 8 \times 1.2 \times 10^{-5+3+4}}{3 \times 4 \times 10^{6-6}}$$

$$= \frac{11.52 \times 10^{2}}{12} = \frac{1152}{12} = 96$$

7. Squares

The square of a number is simply the number multiplied by itself once, and this is true of numbers expressed with powers of ten. The coefficient, or part of the number containing significant digits, is first multiplied by itself in the ordinary way, and the product is then multiplied by a new power of ten obtained by multiplying the original exponent by 2.

With a little practice this becomes easy.

EXAMPLES:

(a) $437^2 = (4.37 \times 10^2)^2$
$= 4.37 \times 4.37 \times 10^4$
$= 19.0969 \times 10^4$
$\cong 19.1 \times 10^4$ (correct to 3 significant digits)
In common notation this is 191,000.

(b) $0.00178^2 = (1.78 \times 10^{-3})^2$
$= 1.78 \times 1.78 \times 10^{-6}$
$= 3.1684 \times 10^{-6}$
$\cong 3.17 \times 10^{-6}$ (correct to 3 significant digits)
In common notation this is 0.00000317.

8. Square Roots

The easy way to use the powers of ten method for calculating square roots requires a little variation from the scientific notation form. *In converting from the common notation, write the coefficient, or part containing the significant digits, as a number between 1 and 100, times an even power of ten. Then take the square root of the coefficient and multiply it by a new power of ten obtained by dividing the original exponent by 2.*

With a little practice this becomes fairly easy, but a small electronic calculator or a slide rule is helpful.

EXAMPLES:

(a) $\sqrt{92800} = \sqrt{9.28 \times 10^4} = 3.0463 \times 10^{4/2}$
$\cong 3.046 \times 10^2$ (correct to 4 significant digits)

In common notation this is $304.6 \cong 305$. Notice that it is hard to estimate the magnitude of the answer when the square root of 92800 is in the common notation, but in the powers of ten notation it is obvious that the square root of 9.28 is a little over three, and this is multiplied by 10^2 or 100, so the correct answer must be a little over 300.

(b) $\sqrt{0.00348} = \sqrt{34.8 \times 10^{-4}} = 5.899 \times 10^{-4/2}$
$= 5.90 \times 10^{-2}$

In common notation this is 0.059.

9. Cubes

The cube of a number is obtained by multiplying the number by itself twice, and this is true of numbers expressed with powers of ten. The coefficient, or part of the number containing significant digits, is first multiplied twice, and the product is then multiplied by a new power of ten obtained by multiplying the original exponent by 3.

With a little practice this become easy although it is rather tedious without a calculator or slide rule.

EXAMPLES:

(a) $56.9^3 = (5.69 \times 10^1)^3 = 184.22 \times 10^3$
$= 1.842 \times 10^5$ (in scientific notation)

In common notation this is 184,200.

(b) $0.0473^3 = (4.73 \times 10^{-2})^3 = 105.82 \times 10^{-6}$
$= 1.058 \times 10^{-4}$ (in scientific notation)

This is 0.0001058 in common notation.

10. Cube Roots

The easy way to use the powers of ten method for calculating cube roots requires a little variation from the scientific notation form. *In converting from the common notation, write the coefficient, or part containing the significant digits, as a number between 1 and 1000 times a power of ten that has 3 as a factor, such as 10^3, 10^6, 10^9 or 10^{-3}, 10^{-6}, or 10^{-9}. Then take the cube root of the coefficient and multiply it by the new power of ten obtained by dividing the original exponent by 3.*

Manual extraction of cube roots is extremely tedious, so it is best to utilize tables or a slide rule.

EXAMPLES:

(a) $\sqrt[3]{4920} = \sqrt[3]{4.92 \times 10^3} = 1.70 \times 10^{3/3}$
$= 1.70 \times 10^1$

In common notation this is 17.

(b) $\sqrt[3]{0.000236} = \sqrt[3]{236 \times 10^{-6}} = 6.18 \times 10^{-6/3}$
$= 6.18 \times 10^{-2}$

And this is 0.0618 in common notation.

A useful fact to remember about cube roots is that the cube root of any number between 1 and 1000 is a number between 1 and 10. For instance, the cube root of 1.030301 is 1.01, and the cube root of 997.00299 is 9.99. And both 1.01 and 9.99 are between 1 and 10.

11. Quick Approximations

Many people find it very difficult to estimate the size to expect in answers to calculations when both large and small numbers occur in the same problem. The powers of ten method is ideal in cases of this kind and for making quick approximations for "ball park" figures that can later be worked out by calculator or slide rule.

To obtain a quick, rough approximation, start out by rewriting the problem entirely in standard form, in which all numbers are basically reduced to numbers between 1 and 10 times the appropriate positive or negative powers of ten. Then rewrite a second time, simplifying the coefficients, or parts containing the significant digits, writing them as the nearest whole numbers. Perform the indicated operations. Then multiply this calculation by the appropriate power of ten.

EXAMPLES:

(a) Problem as stated: $\dfrac{0.959 \times 37.5 \times 427}{185{,}000 \times 0.0637}$

In standard form: $= \dfrac{9.59 \times 10^{-1} \times 3.75 \times 10^{1} \times 4.27 \times 10^{2}}{1.85 \times 10^{5} \times 6.37 \times 10^{-2}}$

$$= \frac{9.59 \times 3.75 \times 4.27 \times 10^{-1+1+2}}{1.85 \times 6.37 \times 10^{5-2}}$$

Rough approximation:

$$= \frac{10 \times 4 \times 4 \times 10^{2}}{2 \times 6 \times 10^{3}}$$

$$= \frac{40}{3} \times 10^{2-3} \text{ (after some cancellations)}$$

$$= 13.3 \times 10^{-1}$$

$$= 1.33$$

Fully computed on electronic calculator: 1.3030664

(b) Problem as stated: $\dfrac{167{,}000 \times 29 \times 4}{30{,}000{,}000 \times 3.14 \times 8.5}$

$$= \frac{1.67 \times 10^{5} \times 2.9 \times 10^{1} \times 4}{3 \times 10^{7} \times 3.14 \times 8.5}$$

$$= \frac{1.67 \times 2.9 \times 4 \times 10^{5+1}}{3 \times 3.14 \times 8.5 \times 10^{7}}$$

$$= \frac{1.67 \times 2.9 \times 4}{3 \times 3.14 \times 8.5} = 10^{-1}$$

Rough approximation: $\dfrac{2 \times 3 \times 4}{3 \times 3 \times 9} \times 10^{-1}$

$$= \frac{8}{27} \times 10^{-1} \cong 0.3 \times 10^{-1}$$

In common notation 0.03

(Fully computed on electronic calculator: 0.0241938.) This approximation, by no means a close one, is still useful for determining the magnitude of the answer.

12. The E Notation

Still another notation is available for expressing powers of ten. This is known as the "E" notation, and is used more by metric countries than in the United States. "E" stands for exponent, and is well suited for computer processing and data transmission because it requires no writing of exponents. Here the *E* stands for the base 10 of our decimal system, and positive or negative exponents, written with plus or minus signs follow the E.

As an example, previous Problem (b) could be written in E notation as: $\frac{1.67 \times 2.9 \times 4 \text{ E} + 5 + 1}{3 \times 3.14 \times 8.5 \text{ E} + 7}$

$$= \frac{1.67 \times 2.9 \times 4 \text{ E} + 6}{3 \times 3.14 \times 8.5 \text{ E} + 7}$$

$$= \frac{1.67 \times 2.9 \times 4}{3 \times 3.14 \times 8.5} \text{ E} - 1 \cong 0.242 \text{ E} - 1,$$

or 0.0242 in common notation

Using the "E" notation, this can be expressed in a single line as $(1.67 \times 2.9 \times 4) \div (3 \times 3.14 \times 8.5)$ E − 1. Better yet, to facilitate operations on a small electronic calculator, this can be restated as $1.67 \times 2.9 \times 4 \div 3 \div 3.14 \div 8.5 \times 0.1$. Just perform each indicated operation in sequence without stopping—first all the multiplications except the last one, then divisions, and finally the multiplication $\times$ 0.1. The final multiplication stands for E −1 and converts the calculation back to common notation.

Calculators and computers produce an enormous number of digits in such cases. This is because they are utterly truthful but gullible, accepting all input data as exact. This is the fallacy of false accuracy. Using data based on instrument readings, an answer rounded to 0.0242 is more realistic than 0.0241938.

Appendix C

ALTERNATING CURRENT FORMULAS

ALTERNATING CURRENT

$$E = IZ = \frac{P}{I\cos\theta} = \sqrt{\frac{PZ}{\cos\theta}}$$

$$I = \frac{E}{Z} = \frac{P}{E\cos\theta} = \sqrt{\frac{P}{Z\cos\theta}}$$

$$P = IE\cos\theta = I^2Z\cos\theta = \frac{E^2\cos\theta}{Z}$$

$$Z = \frac{E}{I} = \frac{P}{I^2\cos\theta} = \frac{E^2\cos\theta}{P}$$

TABLE 1. FORMULAS FOR PURE RESISTIVE CIRCUITS

Alphabetical List of Symbols Used in Formulas

E = Voltage (Volts); E_1, E_2, E_3, etc. = Individual voltages; E_T = Total voltage
I = Current (Amperes); I_1, I_2, I_3, etc. = Individual currents
I_T = Total current
PF = Power factor (Pure number)
R = Resistance (Ohms); R_1, R_2, R_3, etc. = Individual resistances;
R_T = Total resistance

FORMULAS

SERIES RESISTIVE CIRCUITS

$$E_T = E_1 + E_2 + E_3 +, \text{etc.}$$
$$I_T = I_1 = I_2 = I_3, \text{etc.}$$
$$R_T = R_1 + R_2 + R_3 +, \text{etc.}$$

PARALLEL RESISTIVE CIRCUITS

$$E_T = E_1 = E_2 = E_3, \text{etc.}$$
$$I_T = I_1 + I_2 + I_3, \text{etc.}$$
$$R_T = \frac{R_1 \times R_2}{R_1 + R_2}$$
$$R_T = \frac{1}{\frac{1}{R_1} + \frac{1}{R_2} + \frac{1}{R_3} +, \text{etc.}}$$

$$\frac{1}{R_T} = \frac{1}{R_1} + \frac{1}{R_2} + \frac{1}{R_3} +, \text{etc.}$$

POWER FACTOR

$$PF = 1$$

TABLE 2. FORMULAS FOR PURE INDUCTIVE CIRCUITS

Alphabetical List of Symbols Used in Formulas

E = Voltage (Volts); E_P = Voltage in primary; E_S = Voltage in secondary
EFF = Efficiency (Percent)
I = Current (Amperes); I_P = Current in primary; I_S = Current in secondary
L = Inductance (Henrys); L_1, L_2, L_3, etc. = Individual inductors;
L_M = Mutual inductance; L_T = Total inductance
N = Number of turns in coil; N_P = Number of turns in primary;
N_S = Number of turns in secondary
P = Power (Watts); P_{in} = Power in; P_{out} = Power out
R = Resistance of inductor (Ohms); R_L = Load resistance
TC = Time constant (Seconds)

FORMULAS

INDUCTORS IN SERIES

$$L_T = L_1 + L_2 + L_3, \text{etc.}$$

INDUCTORS IN PARALLEL

$$L_T = \frac{1}{\frac{1}{L_1} + \frac{1}{L_2} + \frac{1}{L_3}, \text{etc.}}$$

INDUCTIVE TIME CONSTANT (CURRENT LAGS VOLTAGE)

$$TC = \frac{L}{R}$$

TRANSFORMER TURNS, AND EFFICIENCY

$$\text{Turns ratio} = \frac{N_S}{N_P}$$

$$\frac{E_S}{E_P} = \frac{N_S}{N_P}$$

$$I_S = \frac{N_S}{R_L}$$

$$\frac{I_P}{I_S} = \frac{E_S}{E_P}$$

$$EFF = \frac{P_{in}}{P_{out}} \times 100$$

TABLE 3. FORMULAS FOR PURE CAPACITIVE CIRCUITS (AC AND PULSATING DC ONLY)

Alphabetical List of Symbols Used in Formulas

C = Capacitance (Farads); C_1, C_2, C_3, etc. = Individual capacitors
C_T = Total Capacitance

E = Voltage (Volts)

Q = Electrical charge (Coulombs)

R = Resistance (Ohms)

TC = Time constant (Seconds)

FORMULAS

CAPACITANCE CHARGE

$$Q = CE$$

$$E = \frac{Q}{C}$$

$$C = \frac{Q}{E}$$

SERIES CAPACITIVE CIRCUITS

$$C_T = \frac{C_1C_2}{C_1 + C_2}$$

$$\frac{1}{C_T} = \frac{1}{C_1} + \frac{1}{C_2} + \frac{1}{C_3}\text{, etc.}$$

$$C_T = \frac{1}{\frac{1}{C_1} + \frac{1}{C_2} + \frac{1}{C_3} + \text{, etc.}}$$

PARALLEL CAPACITIVE CIRCUITS

$$C_T = C_1 + C_2 + C_3\text{, etc.}$$

CAPACITIVE TIME CONSTANT (CURRENT LEADS VOLTAGE)

$$TC = R \times C$$

TABLE 4. FORMULAS FOR RL CIRCUITS (COMBINED RESISTANCE AND INDUCTANCE)

Alphabetical List of Symbols Used in Formulas

cos = Cosine (Pure number)

E = Voltage (Volts); E_A = Applied voltage; E_1, E_2, etc. = Individual voltages; E_{X_L} = Inductor voltage; E_R = Voltage across resistance; E_T = Total voltage

f = Frequency (Hertz)

I = Current (Amperes); I_1, I_2, etc. = Individual currents; I_R = Current through resistor; I_T = Total current; I_{X_L} = Inductive current

L = Inductance (Henrys); L_1, L_2, etc. = Individual inductors; L_T = Total inductance

R = Resistance (Ohms); R_1, R_2, etc. = Individual resistors; R_T = Total resistance

TC = Time constant (Seconds)

tan = Tangent (Pure number)

θ (Theta) = Phase angle between applied voltage and current (Degrees)

X = Reactance (Ohms); X_L = Inductive reactance (Ohms); X_{L_1}, X_{L_2}, etc. = Individual inductive reactances; X_{L_T} = Total inductive reactance

Z = Impedance (Ohms)

FORMULAS

INDUCTIVE REACTANCE (Single inductor)

$$X_L = 2\pi fL$$

$$L = \frac{X_L}{2\pi f}$$

$$E = I \times X_L$$

$$X_L = \frac{E}{I}$$

$$I = \frac{E}{X_L}$$

NOTE: $2\pi = 6.28$ (approx.)

SERIES INDUCTIVE REACTANCE

$$X_{L_T} = X_{L_1} + X_{L_2} + X_{L_3} +, \text{etc.}$$

PARALLEL INDUCTIVE REACTANCE

$$X_{L_T} = \frac{X_{L_1} \times X_{L_2}}{X_{L_1} + X_{L_2}}$$

$$\frac{1}{X_{L_T}} = \frac{1}{X_{L_1}} + \frac{1}{X_{L_2}} + \frac{1}{X_{L_3}} +, \text{etc.}$$

$$X_{L_T} = \frac{1}{\frac{1}{X_{L_1}} + \frac{1}{X_{L_2}} + \frac{1}{X_{L_3}}, \text{etc.}}$$

X_L AND RESISTANCE IN SERIES

$$Z = \sqrt{R^2 + (X_L)^2}$$

$$E_A = \sqrt{(E_R)^2 + (E_{X_L})^2}$$

$$\text{TAN } \theta = \frac{X_L}{R}$$

X_L AND RESISTANCE IN PARALLEL

$$I_T = \sqrt{(I_R)^2 + (I_{X_L})^2}$$

$$\text{TAN } \theta = \frac{-I_{X_L}}{I_R}$$

INDUCTIVE TIME CONSTANT

$$TC = \frac{L}{R}$$

TABLE 5. FORMULAS FOR RC CIRCUITS (COMBINED RESISTANCE AND CAPACITANCE)

Alphabetical List of Symbols Used in Formulas

C = Capacitance (Farads); C_1, C_2, etc. = Individual capacitances;
C_T = Total capacitance

cos = Cosine (Pure number)

E = Voltage (Volts); E_A = Applied voltage; E_1, E_2, etc. = Individual voltages;
E_R = Voltage across resistance; E_T = Total voltage;
E_{X_C} = Capacitor voltage

f = Frequency (hertz)

I = Current (Amperes); I_1, I_2, etc. = Individual currents;
I_R = Current through resistor; I_T = Total current;
I_{X_C} = Capacitive current

R = Resistance (Ohms); R_1, R_2, etc. = Individual resistances;
R_T = Total resistance

TC = Time constant (Seconds)

tan = Tangent (Pure number)

θ (Theta) = Phase angle between applied voltage and current (Degrees)

X = Reactance (Ohms); X_C = Capacitive reactance (Ohms);

X_{C_1}, X_{C_2}, etc. = Individual capacitive reactances;
C_T = Total capacitive reactance
Z = Impedance (Ohms)

FORMULAS

CAPACITIVE REACTANCE (SINGLE CAPACITOR)

$$X_C = \frac{1}{2\pi fC} = \frac{E}{I}$$

$$C = \frac{1}{2\pi fX_C}$$

$$E = I \times X_C$$

NOTE: $2\pi = 6.28$ (APPROX.)

$$I = \frac{E}{X_C}$$

SERIES CAPACITIVE REACTANCE

$$X_{C_T} = X_{C_1} + X_{C_2} + X_{C_3} + \text{etc.}$$

PARALLEL CAPACITIVE REACTANCE

$$\frac{1}{X_{C_T}} = \frac{1}{X_{C_1}} + \frac{1}{X_{C_2}} + \text{etc.}$$

$$X_{C_T} = \frac{X_{C_1} \times X_{C_2}}{X_{C_1} + X_{C_2}}$$

$$X_{C_T} = \frac{1}{\frac{1}{X_{C_1}} + \frac{1}{X_{C_2}} + \frac{1}{X_{C_3}}\ \text{. etc.}}$$

X_C AND RESISTANCE IN SERIES

$$Z = \sqrt{R^2 + (X_C)^2}$$

$$E_A = \sqrt{(E_R)^2 + (E_{X_C})^2}$$

$$\text{Cos}\ \theta = \frac{R}{Z}$$

$$\text{Tan}\ \theta = -\frac{X_C}{R}$$

X_C AND RESISTANCE IN PARALLEL

$$I_T = \sqrt{(I_R)^2 + (I_{X_C})^2}$$

$$Z = \frac{E}{I_T}$$

$$\text{Tan}\ \theta = \frac{I_{X_C}}{I_R}$$

CAPACITIVE TIME CONSTANT

$$TC = R \times C$$

TABLE 6. FORMULAS FOR RCL CIRCUITS (COMBINED RESISTANCE, CAPACITANCE, AND INDUCTANCE)

Alphabetical List of Symbols Used in Formulas

C = Capacitance (Farads); C_1, C_2, etc. = Individual Capacitances;
  C = Total Capacitance

cos = Cosine (Pure number)

E = Voltage (Volts); E_A = Applied voltage; E_1, E_2, etc. = Individual voltages;
  E_{X_L} = Inductor voltage; E_R = Voltage across resistance;
  E_T = Total voltage; E_{X_C} = Capacitor voltage; E_{X_L} = Inductor voltage

f = Frequency (Hertz); f_r = resonant frequency

I = Current (Amperes); I_1, I_2, etc. = Individual currents;
  I_R = Current through resistor; I_T = Total current;
  I_{X_C} = Capacitive current; I_{X_L} = Inductive current

L = Inductance (Henrys); L_1, L_2, etc. = Individual inductors;
  L_T = Total inductance

R = Resistance (Ohms); R_1, R_2, etc. = Individual resistors;
  R_T = Total resistance

TC = Time constant (Seconds)

tan = Tangent (Pure number)

θ (Theta) = Phase angle between applied voltage and current (Degrees)

X = Reactance (Ohms); X_C = Capacitive reactance (Ohms);
  X_{C_1}, X_{C_2}, etc. = Individual capacitive reactances;
  X_{C_T} = Total capacitive reactance; X_L = Inductive reactance (Ohms);
  X_{L_1}, X_{L_2}, etc. = Individual inductive reactances;
  X_{L_T} = Total inductive reactance

Z = Impedance (Ohms)

FORMULAS

SERIES RCL CIRCUITS

(Angle is minus (−) if X_C is larger and plus (+) if X_L is larger.)

$$Z = \sqrt{R^2 + (X_L - X_C)^2}$$

$$I = \frac{E}{Z}$$

$$\text{Tan } \theta = \frac{X_L - X_C}{R}$$

PARALLEL RCL CIRCUITS

(Angle is minus (−) if I_{X_L} is larger and plus (+) if I_{X_C} is larger)

$$I_T = \sqrt{(I_R)^2 + (I_{X_L} - I_{X_C})^2}$$

$$Z = \frac{E}{I_T}$$

$$\text{Tan } \theta = \frac{I_{X_L} - I_{X_C}}{I_R}$$

RESONANCE

$$f_r = \frac{1}{2\pi \sqrt{LC}}$$

$$L = \frac{1}{(2\pi)^2 (f_R)^2 C}$$

NOTE: $2\pi = 6.28$ (approx.)
$(2\pi)^2 = 39.5$ (approx.)

$$C = \frac{1}{(2\pi)^2 (f_r)^2 L}$$

TABLE 7. ELECTRICAL POWER CALCULATIONS

Alphabetical List of Symbols Used in Formulas

AC = Alternating current
C = Capacitance (Farads)
cos = Cosine (Pure number)
DC = Direct Current
E = Voltage (Volts)
EFF* = Efficiency (as a decimal fraction); E% = Percentage efficiency
G = Conductance (Mhos)
HP = Horsepower (550 ft-lb per second, or 746 Watts)
I = Current (Amperes)
L = Inductance (Henrys)
P = Power (Watts); P_{in} = Power in; P_{out} = Power out
PF = Power factor (a decimal fraction)
R = Resistance (Ohms)
θ (Theta) = Phase angle (Degrees) between voltage and current waves
X = Reactance (Ohms); X_C = Capacitive reactance; X_L = Inductive reactance
Z = Impedance (Ohms)

*(NOTE: The EFF factor applies to *motors* only. It is usually between 0.8 and 0.9 and is determined by tests. Other formulas in this section are used in other electrical power applications.)

FORMULAS

POWER FACTOR (AC ONLY)

$$PF = \cos\theta = \frac{R}{Z} = \frac{P}{EI}$$

$$PF = \frac{R}{\sqrt{R^2 + X^2}}$$

EFFICIENCY (FOR MOTORS ONLY DETERMINED BY TESTS)

$$EFF = \frac{P_{out}}{P_{in}}$$

TABLE 8. CALCULATIONS OF CYCLES, TIME, VELOCITY AND SINE WAVE VALUES

Alphabetical List of Symbols Used in Formulas

AVE = Average sine wave value
C = Capacitance (Farads)
f = Frequency (hertz)
L = Inductance (Henrys)
λ (Lambda) = Wavelength; λ_{cm} = Wavelength in centimeters;
λ_{ft} = Wavelength in feet; λ_m = Wavelength in meters
π (Pi) = 3.1416
PK = Peak sine wave value; PK to PK = Peak-to-Peak sine wave value
R = Resistance (Ohms)
RMS = Root-Mean-Square sine wave value
T = Time (Seconds)
V = Velocity; V_{cm} = Velocity in centimeters per second;
V_{ft} = Velocity in feet per second; V_m = Velocity in meters per second

FORMULAS

NUMBER OF CYCLES = f × T

CYCLES AND FREQUENCY

$$f = \frac{1}{T}$$

TIME

$$T = \frac{1}{f}$$

VELOCITY

$$V_m = \lambda_m f$$

$$V_{cm} = \lambda_{cm} f$$

$$V_{ft} = \lambda_{ft} f$$

WAVELENGTH

$$\lambda = \frac{V}{f}$$

$$\lambda_m = \frac{3 \times 10^8 \text{ m/sec}}{f}$$

$$\lambda_{cm} = \frac{3 \times 10^{10} \text{ cm/sec}}{f}$$

$$\lambda_{ft} = \frac{1130 \text{ ft/sec}}{f}$$

SINE WAVE VALUES

AVE = 0.637 × PK

RMS = 0.707 × PK

PK = 1.414 × RMS

PK TO PK = 2.828 × RMS

Appendix D

SOLVING ELECTRICAL PROBLEMS WITH TRIGONOMETRY TABLES OR THE SLIDE RULE

The following tables of natural trigonometric functions (sines, cosines, and tangents) are needed to find phase angles and power factors, but they also can be used for solving vector problems by simple trigonometry. The "trig" method is much simpler than calculations using root-mean-squares. Also the "trig" method is readily adaptable to machine calculation and slide rule.

The slide rule best suited for this work is of the decimal trig type, with the *S*, *ST*, and *T* or T_1 and T_2 scales of angles decimally divided. Rules with the T_2 scale are particularly recommended because the tangents of angles larger than 45° can be found on the T_2 scale reading from left to right in the normal manner. Space does not allow a full discussion of plane trigonometry or slide rule operation. However, the best procedures are outlined in the following paragraphs.

It is a good practice to make a small, hand-drawn diagram of the vector triangle to be solved, labeling the parts of the triangle with their electrical symbols. Before doing this, however, examine the four triangles in Fig. 1. These are properly labeled with standard mathematical symbols. Whatever the position of the triangle, small *c* always stands for the hypotenuse opposite the 90° angle of a right triangle, which is labeled as large *C*. Small *a* stands for altitude (whether up or down) opposite the angle of elevation or depression, which is labeled as *A*. Small *b* always stands for base, and the angle opposite the base is labeled as *B*. With this system the functions are:

$$\sin A = \frac{a}{c} \qquad \csc A = \frac{c}{a}$$

$$\cos A = \frac{b}{c} \qquad \sec A = \frac{c}{b}$$

$$\tan A = \frac{a}{b} \qquad \cot A = \frac{b}{a}$$

You will notice that the functions in the righthand column are simply the reciprocals of those in the lefthand column. If it is necessary to know the secant of angle A for any reason, no special tables are needed, because

$$\sec A = \frac{1}{\cos A}.$$

Similarly,

$$\csc A = \frac{1}{\sin A}.$$

Also,

$$\cot A = \frac{\cdot 1}{\tan A}.$$

In Fig. 1 you will notice that angle A shows a minus sign before it when it is in the 4th quadrant. Also you will notice that side a also has a minus side

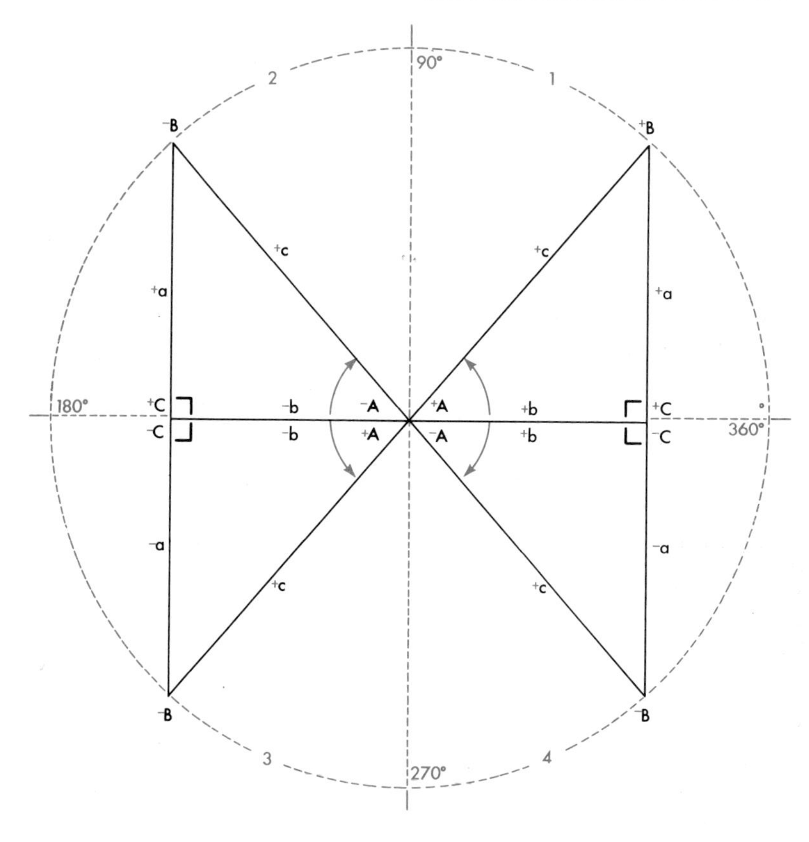

Fig. 1. Right triangles, labeled with standard mathematical symbols, in the four quadrants. Vector triangles for electrical calculations are generally in the 1st or 4th quadrant, and parts are labeled with the appropriate electrical symbols.

when the triangle is in the 4th quadrant. The minus signs do not affect the numerical values; they only indicate the direction.

Fig. 2 shows diagrams of some typical electrical problems, re-labeled with the appropriate electrical symbols. Let us consider one illustrating impedance of an RL circuit and show not only the electrical symbols but their numerical values where these are given.

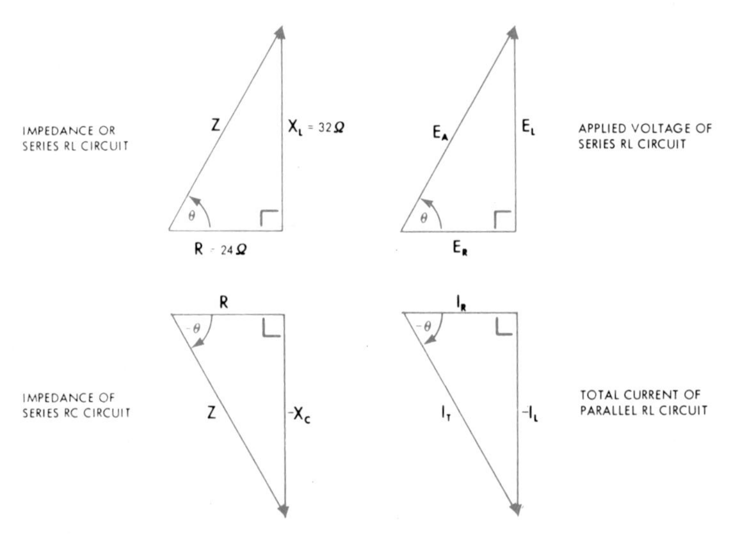

Fig. 2. Four of the many vector diagrams used in electrical work. Parts shown in color are the ones to be found. In each of the cases illustrated the resultant, in whatever electrical unit expressed, is equivalent to c, the hypotenuse, while θ (theta), the phase or displacement angle, is equivalent to angle A in the standard math symbols.

Where $X_L = 32$ and $R = 24$ (both values in ohms) we may want to find the impedance Z, the phase angle θ (theta), and the power factor for this circuit, which is cos θ.

First find the value of θ. This is easy because

$$\tan \theta = \frac{X_L}{R}$$

This is exactly the same as $\tan A = \frac{a}{b}$ in the standard math symbols. Using numbers, $\tan \theta = \frac{32}{24} = 1.3333.$

Because this tangent is larger than 1, the angle is larger than 45°.

From the tables you will not find this number, but you will find the number 1.3319 as the tangent of 53.1° and 1.3367 as the tangent of 53.2°. The exact number will be between 53.1° and 53.2° but nearer to 53.1°. To the nearest tenth of a degree, then, the angle is 53.1°, which is close enough for nearly all electrical calculations.

While still looking at the tables, notice that cos 53.1° = 0.6004, which is almost exactly 0.600. This is the power factor.

Knowing the phase angle θ is 53.1°, it is now very easy to find the impedance Z, using the Law of Sines, that is:

$\frac{c}{\sin C} = \frac{a}{\sin A}$, or, in electrical symbols, $\frac{Z}{\sin 90°} = \frac{X_L}{\sin \theta}$. But the sine of 90° = 1. This simplifies the equation to $Z = \frac{X_L}{\sin \theta}$. Using numbers, $Z = \frac{32}{\sin 53.1°}$. From the tables, sin 53.1° = 0.7997, which is almost exactly 0.8.

$$Z = \frac{32}{.8} = 40 \text{ ohms.}$$

These various operations can be performed on the slide rule in a small fraction of the time required for looking into tables, etc.

Depending on the construction of the slide rule used, there are some differences in the method of finding tangents. Suppose in this example we use a slide rule with a T_2 as well as a T_1 and *ST* scale.

Divide 32 by 24 using the *D* and *C* scales. This gives 1.333 as the tangent (read on the *D* scale below the left index of the *C* scale), but it is not necessary to make any note of this number. At a glance we see that it is larger than 1, know that the angle is therefore larger than 45° and will be read on the T_2 scale.

We move the hairline so it is directly over the left index of the *C* scale, indicating 1.333. The T_2 scale is on the other side of the slide rule, so we flip it over, leaving the hairline in the same position. Still leaving the hairline stationary, we move the slide so that the left index is aligned with the left index of the *D* scale just below the *S* scale. The hairline, which has not been moved, is directly over 53.1 on the T_2 scale, indicating that the angle is 53.1°. We make a note of this for future reference.

If your slide rule has only one *T* scale, an extra step is needed to find tangents larger than 45°. After finding the tangent (in this case 1.333) on the *D* scale you must reset this number on the *CI* scale (an inverse scale with numbers reading from right to left) by moving the hairline. Then read the angle on the *T* scale directly below the hairline on the inverse set of numbers (also reading from right to left) on the *T* scale.

Next we find Z, the impedance, in this manner. Move the hairline to 32 on the *D* scale. Then move the slide until the angle 53.1° on the *S* scale is directly beneath the hairline. Read the answer as 40 (meaning in this case 40 ohms) on the *D* scale beneath the right index of the *S* scale.

To find the power factor, cos θ, we simply align the slide with the left index of the *D* scale, move the hairline along the set of numbers reading *from right to left* on the *S* scale until it is directly above the cosine number 53.1. At this point the hairline is also directly over .600 on the *D* scale. The power factor cos θ is .600.

Although the slide rule does not indicate decimal location, this is no problem in solving triangles, because the magnitude of the quantities is very obvious.

In calculating impedance for RCL circuits, where inductive reactance X_L operates opposite to capacitive reactance X_C, simplify to an equivalent circuit as follows:

1. When values of X_L and X_C are equal they effectively leave a pure resistive circuit at the specified frequency. Voltage and current in this case are in phase, there is no phase angle to consider, no reactance, and no impedance.
2. When X_L is greater than X_C, subtract the value of X_C from X_L. (Subtraction and addition are two operations not performed on the slide rule.) The remainder will be X_L, used for calculating the equivalent RL circuit.
3. When X_C is greater than X_L, subtract X_L from X_C (but not with the slide rule, of course). The remainder will be X_C, used for calculating the equivalent RC circuit.

NATURAL TRIGONOMETRIC FUNCTIONS FOR DECIMAL FRACTIONS OF A DEGREE

Deg.	Sin	Cos	Tan	Cot	Deg.
0.0	.00000	1.0000	.00000	∞	**90.0**
.1	.00175	1.0000	.00175	573.0	.9
.2	.00349	1.0000	.00349	286.5	.8
.3	.00524	1.0000	.00524	191.0	.7
.4	.00698	1.0000	.00698	143.24	.6
.5	.00873	1.0000	.00873	114.59	.5
.6	.01047	0.9999	.01047	95.49	.4
.7	.01222	.9999	.01222	81.85	.3
.8	.01396	.9999	.01396	71.62	.2
.9	.01571	.9999	.01571	63.66	.1
1.0	.01745	0.9998	.01746	57.29	**89.0**
.1	.01920	.9998	.01920	52.08	.9
.2	.02094	.9998	.02095	47.74	.8
.3	.02269	.9997	.02269	44.07	.7
.4	.02443	.9997	.02444	40.92	.6
.5	.02618	.9997	.02619	38.19	.5
.6	.02792	.9996	.02793	35.80	.4
.7	.02967	.9996	.02968	33.69	.3
.8	.03141	.9995	.03143	31.82	.2
.9	.03316	.9995	.03317	30.14	.1
2.0	.03490	0.9994	.03492	28.64	**88.0**
.1	.03664	.9993	.03667	27.27	.9
.2	.03839	.9993	.03842	26.03	.8
.3	.04013	.9992	.04016	24.90	.7
.4	.04188	.9991	.04191	23.86	.6
.5	.04362	.9990	.04366	22.90	.5
.6	.04536	.9990	.04541	22.02	.4
.7	.04711	.9989	.04716	21.20	.3
.8	.04885	.9988	.04891	20.45	.2
.9	.05059	.9987	.05066	19.74	.1
3.0	.05234	0.9986	.05241	19.081	**87.0**
.1	.05408	.9985	.05416	18.464	.9
.2	.05582	.9984	.05591	17.886	.8
.3	.05756	.9983	.05766	17.343	.7
.4	.05931	.9982	.05941	16.832	.6
.5	.06105	.9981	.06116	16.350	.5
.6	.06279	.9980	.06291	15.895	.4
.7	.06453	.9979	.06467	15.464	.3
.8	.06627	.9978	.06642	15.056	.2
.9	.06802	.9977	.06817	14.669	.1
4.0	.06976	0.9976	.06993	14.301	**86.0**
.1	.07150	.9974	.07168	13.951	.9
.2	.07324	.9973	.07344	13.617	.8
.3	.07498	.9972	.07519	13.300	.7
.4	.07672	.9971	.07695	12.996	.6
.5	.07846	.9969	.07870	12.706	.5
.6	.08020	.9968	.08046	12.429	.4
.7	.08194	.9966	.08221	12.163	.3
.8	.08368	.9965	.08397	11.909	.2
.9	.08542	.9963	.08573	11.664	.1
5.0	.08716	0.9962	.08749	11.430	**85.0**
.1	.08889	.9960	.08925	11.205	.9
.2	.09063	.9959	.09101	10.988	.8
.3	.09237	.9957	.09277	10.780	.7
.4	.09411	.9956	.09453	10.579	.6
.5	.09585	.9954	.09629	10.385	.5
.6	.09758	.9952	.09805	10.199	.4
.7	.09932	.9951	.09981	10.019	.3
.8	.10106	.9949	.10158	9.845	.2
.9	.10279	.9947	.10334	9.677	.1
6.0	.10453	0.9945	.10510	9.514	**84.0**
Deg.	Cos	Sin	Cot	Tan	Deg.

Deg.	Sin	Cos	Tan	Cot	Deg.
6.0	.10453	0.9945	.10510	9.514	**84.0**
.1	.10626	.9943	.10687	9.357	.9
.2	.10800	.9942	.10863	9.205	.8
.3	.10973	.9940	.11040	9.058	.7
.4	.11147	.9938	.11217	8.915	.6
.5	.11320	.9936	.11394	8.777	.5
.6	.11494	.9934	.11570	8.643	.4
.7	.11667	.9932	.11747	8.513	.3
.8	.11840	.9930	.11924	8.386	.2
.9	.12014	.9928	.12101	8.264	.1
7.0	.12187	0.9925	.12278	8.144	**83.0**
.1	.12360	.9923	.12456	8.028	.9
.2	.12533	.9921	.12633	7.916	.8
.3	.12706	.9919	.12810	7.806	.7
.4	.12880	.9917	.12988	7.700	.6
.5	.13053	.9914	.13165	7.596	.5
.6	.13226	.9912	.13343	7.495	.4
.7	.13399	.9910	.13521	7.396	.3
.8	.13572	.9907	.13698	7.300	.2
.9	.13744	.9905	.13876	7.207	.1
8.0	.13917	0.9903	.14054	7.115	**82.0**
.1	.14090	.9900	.14232	7.026	.9
.2	.14263	.9898	.14410	6.940	.8
.3	.14436	.9895	.14588	6.855	.7
.4	.14608	.9893	.14767	6.772	.6
.5	.14781	.9890	.14945	6.691	.5
.6	.14954	.9888	.15124	6.612	.4
.7	.15126	.9885	.15302	6.535	.3
.8	.15299	.9882	.15481	6.460	.2
.9	.15471	.9880	.15660	6.386	.1
9.0	.15643	0.9877	.15838	6.314	**81.0**
.1	.15816	.9874	.16017	6.243	.9
.2	.15988	.9871	.16196	6.174	.8
.3	.16160	.9869	.16376	6.107	.7
.4	.16333	.9866	.16555	6.041	.6
.5	.16505	.9863	.16734	5.976	.5
.6	.16677	.9860	.16914	5.912	.4
.7	.16849	.9857	.17093	5.850	.3
.8	.17021	.9854	.17273	5.789	.2
.9	.17193	.9851	.17453	5.730	.1
10.0	.1736	0.9848	.1763	5.671	**80.0**
.1	.1754	.9845	.1781	5.614	.9
.2	.1771	.9842	.1799	5.558	.8
.3	.1788	.9839	.1817	5.503	.7
.4	.1805	.9836	.1835	5.449	.6
.5	.1822	.9833	.1853	5.396	.5
.6	.1840	.9829	.1871	5.343	.4
.7	.1857	.9826	.1890	5.292	.3
.8	.1874	.9823	.1908	5.242	.2
.9	.1891	.9820	.1926	5.193	.1
11.0	.1908	0.9816	.1944	5.145	**79.0**
.1	.1925	.9813	.1962	5.097	.9
.2	.1942	.9810	.1980	5.050	.8
.3	.1959	.9806	.1998	5.005	.7
.4	.1977	.9803	.2016	4.959	.6
.5	.1994	.9799	.2035	4.915	.5
.6	.2011	.9796	.2053	4.872	.4
.7	.2028	.9792	.2071	4.829	.3
.8	.2045	.9789	.2089	4.787	.2
.9	.2062	.9785	.2107	4.745	.1
12.0	.2079	0.9781	.2126	4.705	**78.0**
Deg.	Cos	Sin	Cot	Tan	Deg.

NATURAL TRIGONOMETRIC FUNCTIONS FOR DECIMAL FRACTIONS OF A DEGREE

Deg.	Sin	Cos	Tan	Cot	Deg.	Deg.	Sin	Cos	Tan	Cot	Deg.
12.0	0.2079	0.9781	0.2126	4.705	**78.0**	**18.0**	0.3090	0.9511	0.3249	3.078	**72.0**
.1	.2096	.9778	.2144	4.665	.9	.1	.3107	.9505	.3269	3.060	.9
.2	.2113	.9774	.2162	4.625	.8	.2	.3123	.9500	.3288	3.042	.8
.3	.2130	.9770	.2180	4.586	.7	.3	.3140	.9494	.3307	3.024	.7
.4	.2147	.9767	.2199	4.548	.6	.4	.3156	.9489	.3327	3.006	.6
.5	.2164	.9763	.2217	4.511	.5	.5	.3173	.9483	.3346	2.989	.5
.6	.2181	.9759	.2235	4.474	.4	.6	.3190	.9478	.3365	2.971	.4
.7	.2198	.9755	.2254	4.437	.3	.7	.3206	.9472	.3385	2.954	.3
.8	.2215	.9751	.2272	4.402	.2	.8	.3223	.9466	.3404	2.937	.2
.9	.2233	.9748	.2290	4.366	.1	.9	.3239	.9461	.3424	2.921	.1
13.0	0.2250	0.9744	0.2309	4.331	**77.0**	**19.0**	0.3256	0.9455	0.3443	2.904	**71.0**
.1	.2267	.9740	.2327	4.297	.9	.1	.3272	.9449	.3463	2.888	.9
.2	.2284	.9736	.2345	4.264	.8	.2	.3289	.9444	.3482	2.872	.8
.3	.2300	.9732	.2364	4.230	.7	.3	.3305	.9438	.3502	2.856	.7
.4	.2317	.9728	.2382	4.198	.6	.4	.3322	.9432	.3522	2.840	.6
.5	.2334	.9724	.2401	4.165	.5	.5	.3338	.9426	.3541	2.824	.5
.6	.2351	.9720	.2419	4.134	.4	.6	.3355	.9421	.3561	2.808	.4
.7	.2368	.9715	.2438	4.102	.3	.7	.3371	.9415	.3581	2.793	.3
.8	.2385	.9711	.2456	4.071	.2	.8	.3387	.9409	.3600	2.778	.2
.9	.2402	.9707	.2475	4.041	.1	.9	.3404	.9403	.3620	2.762	.1
14.0	0.2419	0.9703	0.2493	4.011	**76.0**	**20.0**	0.3420	0.9397	0.3640	2.747	**70.0**
.1	.2436	.9699	.2512	3.981	.9	.1	.3437	.9391	.3659	2.733	.9
.2	.2453	.9694	.2530	3.952	.8	.2	.3453	.9385	.3679	2.718	.8
.3	.2470	.9690	.2549	3.923	.7	.3	.3469	.9379	.3699	2.703	.7
.4	.2487	.9686	.2568	3.895	.6	.4	.3486	.9373	.3719	2.689	.6
.5	.2504	.9681	.2586	3.867	.5	.5	.3502	.9367	.3739	2.675	.5
.6	.2521	.9677	.2605	3.839	.4	.6	.3518	.9361	.3759	2.660	.4
.7	.2538	.9673	.2623	3.812	.3	.7	.3535	.9354	.3779	2.646	.3
.8	.2554	.9668	.2642	3.785	.2	.8	.3551	.9348	.3799	2.633	.2
.9	.2571	.9664	.2661	3.758	.1	.9	.3567	.9342	.3819	2.619	.1
15.0	0.2588	0.9659	0.2679	3.732	**75.0**	**21.0**	0.3584	0.9336	0.3839	2.605	**69.0**
.1	.2605	.9655	.2698	3.706	.9	.1	.3600	.9330	.3859	2.592	.9
.2	.2622	.9650	.2717	3.681	.8	.2	.3616	.9323	.3879	2.578	.8
.3	.2639	.9646	.2736	3.655	.7	.3	.3633	.9317	.3899	2.565	.7
.4	.2656	.9641	.2754	3.630	.6	.4	.3649	.9311	.3919	2.552	.6
.5	.2672	.9636	.2773	3.606	.5	.5	.3665	.9304	.3939	2.539	.5
.6	.2689	.9632	.2792	3.582	.4	.6	.3681	.9298	.3959	2.526	.4
.7	.2706	.9627	.2811	3.558	.3	.7	.3697	.9291	.3979	2.513	.3
.8	.2723	.9622	.2830	3.534	.2	.8	.3714	.9285	.4000	2.500	.2
.9	.2740	.9617	.2849	3.511	.1	.9	.3730	.9278	.4020	2.488	.1
16.0	0.2756	0.9613	0.2867	3.487	**74.0**	**22.0**	0.3746	0.9272	0.4040	2.475	**68.0**
.1	.2773	.9608	.2886	3.465	.9	.1	.3762	.9265	.4061	2.463	.9
.2	.2790	.9603	.2905	3.442	.8	.2	.3778	.9259	.4081	2.450	.8
.3	.2807	.9598	.2924	3.420	.7	.3	.3795	.9252	.4101	2.438	.7
.4	.2823	.9593	.2943	3.398	.6	.4	.3811	.9245	.4122	2.426	.6
.5	.2840	.9588	.2962	3.376	.5	.5	.3827	.9239	.4142	2.414	.5
.6	.2857	.9583	.2981	3.354	.4	.6	.3843	.9232	.4163	2.402	.4
.7	.2874	.9578	.3000	3.333	.3	.7	.3859	.9225	.4183	2.391	.3
.8	.2890	.9573	.3019	3.312	.2	.8	.3875	.9219	.4204	2.379	.2
.9	.2907	.9568	.3038	3.291	.1	.9	.3891	.9212	.4224	2.367	.1
17.0	0.2924	0.9563	0.3057	3.271	**73.0**	**23.0**	0.3907	0.9205	0.4245	2.356	**67.0**
.1	.2940	.9558	.3076	3.251	.9	.1	.3923	.9198	.4265	2.344	.9
.2	.2957	.9553	.3096	3.230	.8	.2	.3939	.9191	.4286	2.333	.8
.3	.2974	.9548	.3115	3.211	.7	.3	.3955	.9184	.4307	2.322	.7
.4	.2990	.9542	.3134	3.191	.6	.4	.3971	.9178	.4327	2.311	.6
.5	.3007	.9537	.3153	3.172	.5	.5	.3987	.9171	.4348	2.300	.5
.6	.3024	.9532	.3172	3.152	.4	.6	.4003	.9164	.4369	2.289	.4
.7	.3040	.9527	.3191	3.133	.3	.7	.4019	.9157	.4390	2.278	.3
.8	.3057	.9521	.3211	3.115	.2	.8	.4035	.9150	.4411	2.267	.2
.9	.3074	.9516	.3230	3.096	.1	.9	.4051	.9143	.4431	2.257	.1
18.0	0.3090	0.9511	0.3249	3.078	**72.0**	**24.0**	0.4067	0.9135	0.4452	2.246	**66.0**
Deg.	Cos	Sin	Cot	Tan	Deg.	Deg.	Cos	Sin	Cot	Tan	Deg.

NATURAL TRIGONOMETRIC FUNCTIONS FOR DECIMAL FRACTIONS OF A DEGREE

Deg.	Sin	Cos	Tan	Cot	Deg.	Deg.	Sin	Cos	Tan	Cot	Deg.
24.0	0.4067	0.9135	0.4452	2.246	**66.0**	**30.0**	0.5000	0.8660	0.5774	1.7321	**60.0**
.1	.4083	.9128	.4473	2.236	.9	1	.5015	.8652	.5797	1.7251	.9
.2	.4099	.9121	.4494	2.225	8	2	.5030	.8643	.5820	1.7182	.8
.3	.4115	.9114	.4515	2.215	7	3	.5045	.8634	.5844	1.7113	.7
.4	.4131	.9107	.4536	2.204	6	4	.5060	.8625	.5867	1.7045	.6
.5	.4147	.9100	.4557	2.194	5	5	.5075	.8616	.5890	1.6977	.5
.6	.4163	.9092	.4578	2.184	4	6	.5090	.8607	.5914	1.6909	.4
.7	.4179	.9085	.4599	2.174	3	7	.5105	.8599	.5938	1.6842	.3
.8	.4195	.9078	.4621	2.164	2	8	.5120	.8590	.5961	1.6775	.2
.9	.4210	.9070	.4642	2.154	1	9	.5135	.8581	.5985	1.6709	.1
25.0	0.4226	0.9063	0.4663	2.145	**65.0**	**31.0**	0.5150	0.8572	0.6009	1.6643	**59.0**
1	.4242	.9056	.4684	2.135	9	1	.5165	.8563	.6032	1.6577	.9
2	.4258	.9048	.4706	2.125	8	2	.5180	.8554	.6056	1.6512	.8
3	.4274	.9041	.4727	2.116	7	3	.5195	.8545	.6080	1.6447	.7
4	.4289	.9033	.4748	2.106	6	4	.5210	.8536	.6104	1.6383	.6
.5	.4305	.9026	.4770	2.097	5	5	.5225	.8526	.6128	1.6319	.5
6	.4321	.9018	.4791	2.087	4	6	.5240	.8517	.6152	1.6255	.4
7	.4337	.9011	.4813	2.078	3	7	.5255	.8508	.6176	1.6191	.3
8	.4352	.9003	.4834	2.069	2	8	.5270	.8499	.6200	1.6128	.2
9	.4368	.8996	.4856	2.059	1	9	.5284	.8490	.6224	1.6066	.1
26.0	0.4384	0.8988	0.4877	2.050	**64.0**	**32.0**	0.5299	0.8480	0.6249	1.6003	**58.0**
.1	.4399	.8980	.4899	2.041	9	1	.5314	.8471	.6273	1.5941	.9
.2	.4415	.8973	.4921	2.032	8	2	.5329	.8462	.6297	1.5880	.8
.3	.4431	.8965	.4942	2.023	7	3	.5344	.8453	.6322	1.5818	.7
.4	.4446	.8957	.4964	2.014	6	4	.5358	.8443	.6346	1.5757	.6
.5	.4462	.8949	.4986	2.006	5	5	.5373	.8434	.6371	1.5697	.5
.6	.4478	.8942	.5008	1.997	4	6	.5388	.8425	.6395	1.5637	.4
.7	.4493	.8934	.5029	1.988	3	7	.5402	.8415	.6420	1.5577	.3
.8	.4509	.8926	.5051	1.980	2	8	.5417	.8406	.6445	1.5517	.2
.9	.4524	.8918	.5073	1.971	1	9	.5432	.8396	.6469	1.5458	.1
27.0	0.4540	0.8910	0.5095	1.963	**63.0**	**33.0**	0.5446	0.8387	0.6494	1.5399	**57.0**
.1	.4555	.8902	.5117	1.954	9	1	.5461	.8377	.6519	1.5340	.9
.2	.4571	.8894	.5139	1.946	8	2	.5476	.8368	.6544	1.5282	.8
.3	.4586	.8886	.5161	1.937	7	3	.5490	.8358	.6569	1.5224	.7
.4	.4602	.8878	.5184	1.929	6	4	.5505	.8348	.6594	1.5166	.6
.5	.4617	.8870	.5206	1.921	5	5	.5519	.8339	.6619	1.5108	.5
.6	.4633	.8862	.5228	1.913	4	6	.5534	.8329	.6644	1.5051	.4
.7	.4648	.8854	.5250	1.905	3	7	.5548	.8320	.6669	1.4994	.3
.8	.4664	.8846	.5272	1.897	2	8	.5563	.8310	.6694	1.4938	.2
.9	.4679	.8838	.5295	1.889	1	9	.5577	.8300	.6720	1.4882	.1
28.0	0.4695	0.8829	0.5317	1.881	**62.0**	**34.0**	0.5592	0.8290	0.6745	1.4826	**56.0**
.1	.4710	.8821	.5340	1.873	9	1	.5606	.8281	.6771	1.4770	.9
.2	.4726	.8813	.5362	1.865	8	2	.5621	.8271	.6796	1.4715	.8
.3	.4741	.8805	.5384	1.857	7	3	.5635	.8261	.6822	1.4659	.7
.4	.4756	.8796	.5407	1.849	6	4	.5650	.8251	.6847	1.4605	.6
.5	.4772	.8788	.5430	1.842	5	5	.5664	.8241	.6873	1.4550	.5
.6	.4787	.8780	.5452	1.834	4	6	.5678	.8231	.6899	1.4496	.4
.7	.4802	.8771	.5475	1.827	3	7	.5693	.8221	.6924	1.4442	.3
.8	.4818	.8763	.5498	1.819	2	8	.5707	.8211	.6950	1.4388	.2
.9	.4833	.8755	.5520	1.811	1	9	.5721	.8202	.6976	1.4335	.1
29.0	0.4848	0.8746	0.5543	1.804	**61.0**	**35.0**	0.5736	0.8192	0.7002	1.4281	**55.0**
.1	.4863	.8738	.5566	1.797	9	1	.5750	.8181	.7028	1.4229	.9
.2	.4879	.8729	.5589	1.789	8	2	.5764	.8171	.7054	1.4176	.8
.3	.4894	.8721	.5612	1.782	7	3	.5779	.8161	.7080	1.4124	.7
.4	.4909	.8712	.5635	1.775	6	4	.5793	.8151	.7107	1.4071	.6
.5	.4924	.8704	.5658	1.767	5	5	.5807	.8141	.7133	1.4019	.5
.6	.4939	.8695	.5681	1.760	4	6	.5821	.8131	.7159	1.3968	.4
.7	.4955	.8686	.5704	1.753	3	7	.5835	.8121	.7186	1.3916	.3
.8	.4970	.8678	.5727	1.746	2	8	.5850	.8111	.7212	1.3865	.2
.9	.4985	.8669	.5750	1.739	1	9	.5864	.8100	.7239	1.3814	.1
30.0	0.5000	0.8660	0.5774	1.732	**60.0**	**36.0**	0.5878	0.8090	0.7265	1.3764	**54.0**
Deg.	Cos	Sin	Cot	Tan	Deg.	Deg.	Cos	Sin	Cot	Tan	Deg.

NATURAL TRIGONOMETRIC FUNCTIONS FOR DECIMAL FRACTIONS OF A DEGREE

Deg.	Sin	Cos	Tan	Cot	Deg.	Deg.	Sin	Cos	Tan	Cot	Deg.
36.0	0.5878	0.8090	0.7265	1.3764	**54.0**	**40.5**	0.6494	0.7604	0.8541	1.1708	**49.5**
.1	.5892	.8080	.7292	1.3713	.9	.6	.6508	.7593	.8571	1.1667	.4
.2	.5906	.8070	.7319	1.3663	.8	.7	.6521	.7581	.8601	1.1626	3
.3	.5920	.8059	.7346	1.3613	.7	.8	.6534	.7570	.8632	1.1585	2
.4	.5934	.8049	.7373	1.3564	.6	.9	.6547	.7559	.8662	1.1544	.1
.5	.5948	.8039	.7400	1.3514	.5	**41.0**	0.6561	0.7547	0.8693	1.1504	**49.0**
.6	.5962	.8028	.7427	1.3465	.4	.1	.6574	.7536	.8724	1.1463	.9
.7	.5976	.8018	.7454	1.3416	.3	.2	.6587	.7524	.8754	1.1423	.8
.8	.5990	.8007	.7481	1.3367	.2	.3	.6600	.7513	.8785	1.1383	.7
.9	.6004	.7997	.7508	1.3319	.1	.4	.6613	.7501	.8816	1.1343	.6
37.0	0.6018	0.7986	0.7536	1.3270	**53.0**	.5	.6626	.7490	.8847	1.1303	.5
.1	.6032	.7976	.7563	1.3222	.9	.6	.6639	.7478	.8878	1.1263	.4
.2	.6046	.7965	.7590	1.3175	.8	.7	.6652	.7466	.8910	1.1224	.3
.3	.6060	.7955	.7618	1.3127	.7	.8	.6665	.7455	.8941	1.1184	2
.4	.6074	.7944	.7646	1.3079	.6	.9	.6678	.7443	.8972	1.1145	1
.5	.6088	.7934	.7673	1.3032	.5	**42.0**	0.6691	0.7431	0.9004	1.1106	**48.0**
.6	.6101	.7923	.7701	1.2985	.4	.1	.6704	.7420	.9036	1.1067	9
.7	.6115	.7912	.7729	1.2938	.3	.2	.6717	.7408	.9067	1.1028	8
.8	.6129	.7902	.7757	1.2892	.2	.3	.6730	.7396	.9099	1.0990	.7
.9	.6143	.7891	.7785	1.2846	.1	.4	.6743	.7385	.9131	1.0951	6
38.0	0.6157	0.7880	0.7813	1.2799	**52.0**	.5	.6756	.7373	.9163	1.0913	.5
.1	.6170	.7869	.7841	1.2753	.9	.6	.6769	.7361	.9195	1.0875	.4
.2	.6184	.7859	.7869	1.2708	.8	.7	.6782	.7349	.9228	1.0837	.3
.3	.6198	.7848	.7898	1.2662	.7	.8	.6794	.7337	.9260	1.0799	.2
.4	.6211	.7837	.7926	1.2617	.6	.9	.6807	.7325	.9293	1.0761	.1
.5	.6225	.7826	.7954	1.2572	.5	**43.0**	0.6820	0.7314	0.9325	1.0724	**47.0**
.6	.6239	.7815	.7983	1.2527	.4	.1	.6833	.7302	.9358	1.0686	.9
.7	.6252	.7804	.8012	1.2482	3	.2	.6845	.7290	.9391	1.0649	.8
.8	.6266	.7793	.8040	1.2437	.2	.3	.6858	.7278	.9424	1.0612	.7
.9	.6280	.7782	.8069	1.2393	.1	.4	.6871	.7266	.9457	1.0575	.6
39.0	0.6293	0.7771	0.8098	1.2349	**51.0**	.5	.6884	.7254	.9490	1.0538	.5
.1	.6307	.7760	.8127	1.2305	.9	.6	.6896	.7242	.9523	1.0501	.4
.2	.6320	.7749	.8156	1.2261	8	.7	.6909	.7230	.9556	1.0464	.3
.3	.6334	.7738	.8185	1.2218	7	.8	.6921	.7218	.9590	1.0428	.2
.4	.6347	.7727	.8214	1.2174	.6	.9	.6934	.7206	.9623	1.0392	.1
.5	.6361	.7716	.8243	1.2131	.5	**44.0**	0.6947	0.7193	0.9657	1.0355	**46.0**
.6	.6374	.7705	.8273	1.2088	.4	.1	.6959	.7181	.9691	1.0319	.9
.7	.6388	.7694	.8302	1.2045	.3	.2	.6972	.7169	.9725	1.0283	.8
.8	.6401	.7683	.8332	1.2002	.2	.3	.6984	.7157	.9759	1.0247	.7
.9	.6414	.7672	.8361	1.1960	.1	.4	.6997	.7145	.9793	1.0212	.6
40.0	0.6428	0.7660	0.8391	1.1918	**50.0**	.5	.7009	.7133	.9827	1.0176	.5
.1	.6441	.7649	.8421	1.1875	.9	.6	.7022	.7120	.9861	1.0141	.4
.2	.6455	.7638	.8451	1.1833	.8	.7	.7034	.7108	.9896	1.0105	.3
3	.6468	.7627	.8481	1.1792	.7	.8	.7046	.7096	.9930	1.0070	.2
.4	.6481	.7615	.8511	1.1750	.6	.9	.7059	.7083	.9965	1.0035	.1
40.5	0.6494	0.7604	0.8541	1.1708	**49.5**	**45.0**	0.7071	0.7071	1.0000	1.0000	**45.0**
Deg.	Cos	Sin	Cot	Tan	Deg.	Deg.	Cos	Sin	Cot	Tan	Deg.

Appendix E

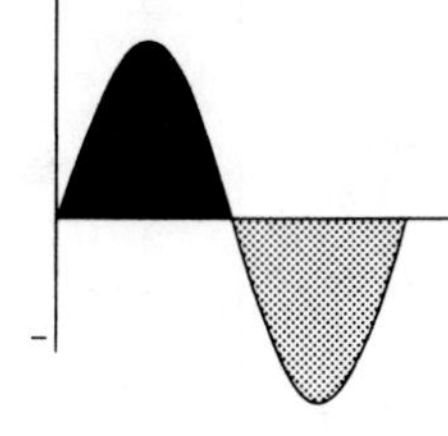

Squares, Cubes, Roots, and Reciprocals of Numbers from 1 to 100

See pages 126 and 127 for tables.

SQUARES, CUBES, SQUARE ROOTS, CUBE ROOTS, AND RECIPROCALS OF NUMBERS FROM 1 TO 50

No.	Square	Cube	Sq. Root	Cube Root	Reciprocal
1	1	1	1.00000	1.00000	1.0000000
2	4	8	1.41421	1.25992	0.5000000
3	9	27	1.73205	1.44225	0.3333333
4	16	64	2.00000	1.58740	0.2500000
5	25	125	2.23607	1.70998	0.2000000
6	36	216	2.44949	1.81712	0.1666667
7	49	343	2.64575	1.91293	0.1428571
8	64	512	2.82843	2.00000	0.1250000
9	81	729	3.00000	2.08008	0.1111111
10	100	1,000	3.16228	2.15443	0.1000000
11	121	1,331	3.31662	2.22398	0.0909091
12	144	1,728	3.46410	2.28943	0.0833333
13	169	2,197	3.60555	2.35133	0.0769231
14	196	2,744	3.74166	2.41014	0.0714286
15	225	3,375	3.87298	2.46621	0.0666667
16	256	4,096	4.00000	2.51984	0.0625000
17	289	4,913	4.12311	2.57128	0.0588235
18	324	5,832	4.24264	2.62074	0.0555556
19	361	6,859	4.35890	2.66840	0.0526316
20	400	8,000	4.47214	2.71442	0.0500000
21	441	9,261	4.58258	2.75892	0.0476190
22	484	10,648	4.69042	2.80204	0.0454545
23	529	12,167	4.79583	2.84387	0.0434783
24	576	13,824	4.89898	2.88450	0.0416667
25	625	15,625	5.00000	2.92402	0.0400000
26	676	17,576	5.09902	2.96250	0.0384615
27	729	19,683	5.19615	3.00000	0.0370370
28	784	21,952	5.29150	3.03659	0.0357143
29	841	24,389	5.38516	3.07232	0.0344828
30	900	27,000	5.47723	3.10723	0.0333333
31	961	29,791	5.56776	3.14138	0.0322581
32	1,024	32,768	5.65685	3.17480	0.0312500
33	1,089	35,937	5.74456	3.20753	0.0303030
34	1,156	39,304	5.83095	3.23961	0.0294118
35	1,225	42,875	5.91608	3.27107	0.0285714
36	1,296	46,656	6.00000	3.30193	0.0277778
37	1,369	50,653	6.08276	3.33222	0.0270270
38	1,444	54,872	6.16441	3.36198	0.0263158
39	1,521	59,319	6.24500	3.39121	0.0256410
40	1,600	64,000	6.32456	3.41995	0.0250000
41	1,681	68,921	6.40312	3.44822	0.0243902
42	1,764	74,088	6.48074	3.47603	0.0238095
43	1,849	79,507	6.55744	3.50340	0.0232558
44	1,936	85,184	6.63325	3.53035	0.0227273
45	2,025	91,125	6.70820	3.55689	0.0222222
46	2,116	97,336	6.78233	3.58305	0.0217391
47	2,209	103,823	6.85565	3.60883	0.0212766
48	2,304	110,592	6.92820	3.63424	0.0208333
49	2,401	117,649	7.00000	3.65931	0.0204082
50	2,500	125,000	7.07107	3.68403	0.0200000
No.	Square	Cube	Sq. Root	Cube Root	Reciprocal

SQUARES, CUBES, SQUARE ROOTS, CUBE ROOTS, AND RECIPROCALS OF NUMBERS FROM 51 TO 100

No.	Square	Cube	Sq. Root	Cube Root	Reciprocal
51	2,601	132,651	7.14143	3.70843	0.0196078
52	2,704	140,608	7.21110	3.73251	0.0192308
53	2,809	148,877	7.28011	3.75629	0.0188679
54	2,916	157,464	7.34847	3.77976	0.0185185
55	3,025	166,375	7.41620	3.80295	0.0181818
56	3,136	175,616	7.48331	3.82586	0.0178571
57	3,249	185,193	7.54983	3.84850	0.0175439
58	3,364	195,112	7.61577	3.87088	0.0172414
59	3,481	205,379	7.68115	3.89300	0.0169492
60	3,600	216,000	7.74597	3.91487	0.0166667
61	3,721	226,981	7.81025	3.93650	0.0163934
62	3,844	238,328	7.87401	3.95789	0.0161290
63	3,969	250,047	7.93725	3.97906	0.0158730
64	4,096	262,144	8.00000	4.00000	0.0156250
65	4,225	274,625	8.06226	4.02073	0.0153846
66	4,356	287,496	8.12404	4.04124	0.0151515
67	4,489	300,763	8.18535	4.06155	0.0149254
68	4,624	314,432	8.24621	4.08166	0.0147059
69	4,761	328,509	8.30662	4.10157	0.0144928
70	4,900	343,000	8.36660	4.12129	0.0142857
71	5,041	357,911	8.42615	4.14082	0.0140845
72	5,184	373,248	8.48528	4.16017	0.0138889
73	5,329	389,017	8.54400	4.17934	0.0136986
74	5,476	405,224	8.60233	4.19834	0.0135135
75	5,625	421,875	8.66025	4.21716	0.0133333
76	5,776	438,976	8.71780	4.23582	0.0131579
77	5,929	456,533	8.77496	4.25432	0.0129870
78	6,084	474,552	8.83176	4.27266	0.0128205
79	6,241	493,039	8.88819	4.29084	0.0126582
80	6,400	512,000	8.94427	4.30887	0.0125000
81	6,561	531,441	9.00000	4.32675	0.0123457
82	6,724	551,368	9.05539	4.34448	0.0121951
83	6,889	571,787	9.11043	4.36207	0.0120482
84	7,056	592,704	9.16515	4.37952	0.0119048
85	7,225	614,125	9.21954	4.39683	0.0117647
86	7,396	636,056	9.27362	4.41400	0.0116279
87	7,569	658,503	9.32738	4.43105	0.0114943
88	7,744	681,472	9.38083	4.44797	0.0113636
89	7,921	704,969	9.43398	4.46475	0.0112360
90	8,100	729,000	9.48683	4.48140	0.0111111
91	8,281	753,571	9.53939	4.49794	0.0109890
92	8,464	778,688	9.59166	4.51436	0.0108696
93	8,649	804,357	9.64365	4.53065	0.0107527
94	8,836	830,584	9.69536	4.54684	0.0106383
95	9,025	857,375	9.74679	4.56290	0.0105263
96	9,216	884,736	9.79796	4.57886	0.0104167
97	9,409	912,673	9.84886	4.59470	0.0103093
98	9,604	941,192	9.89949	4.61044	0.0102041
99	9,801	970,299	9.94987	4.62607	0.0101010
100	10,000	1,000,000	10.00000	4.64159	0.0100000
No.	**Square**	**Cube**	**Sq. Root**	**Cube Root**	**Reciprocal**

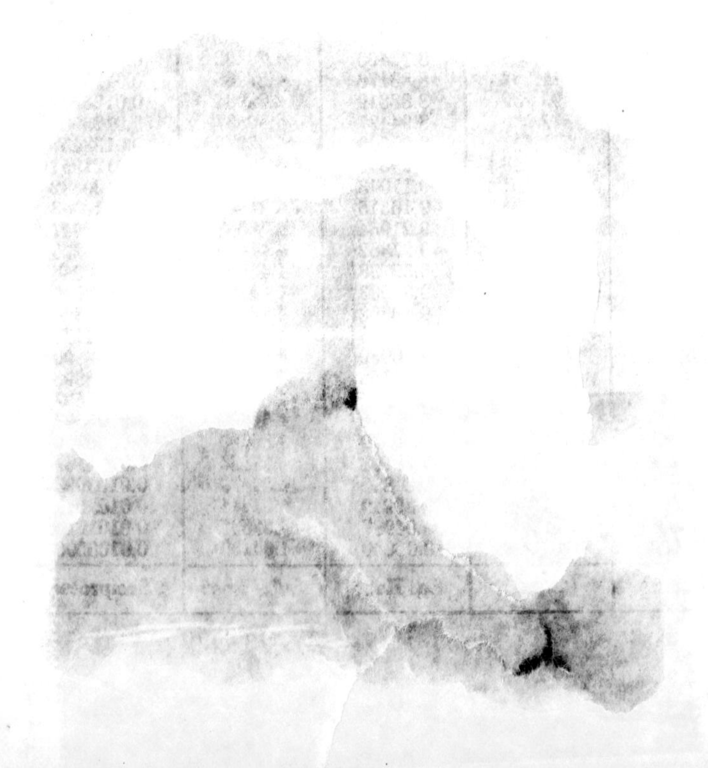